COLOR
IN REPRODUCTION

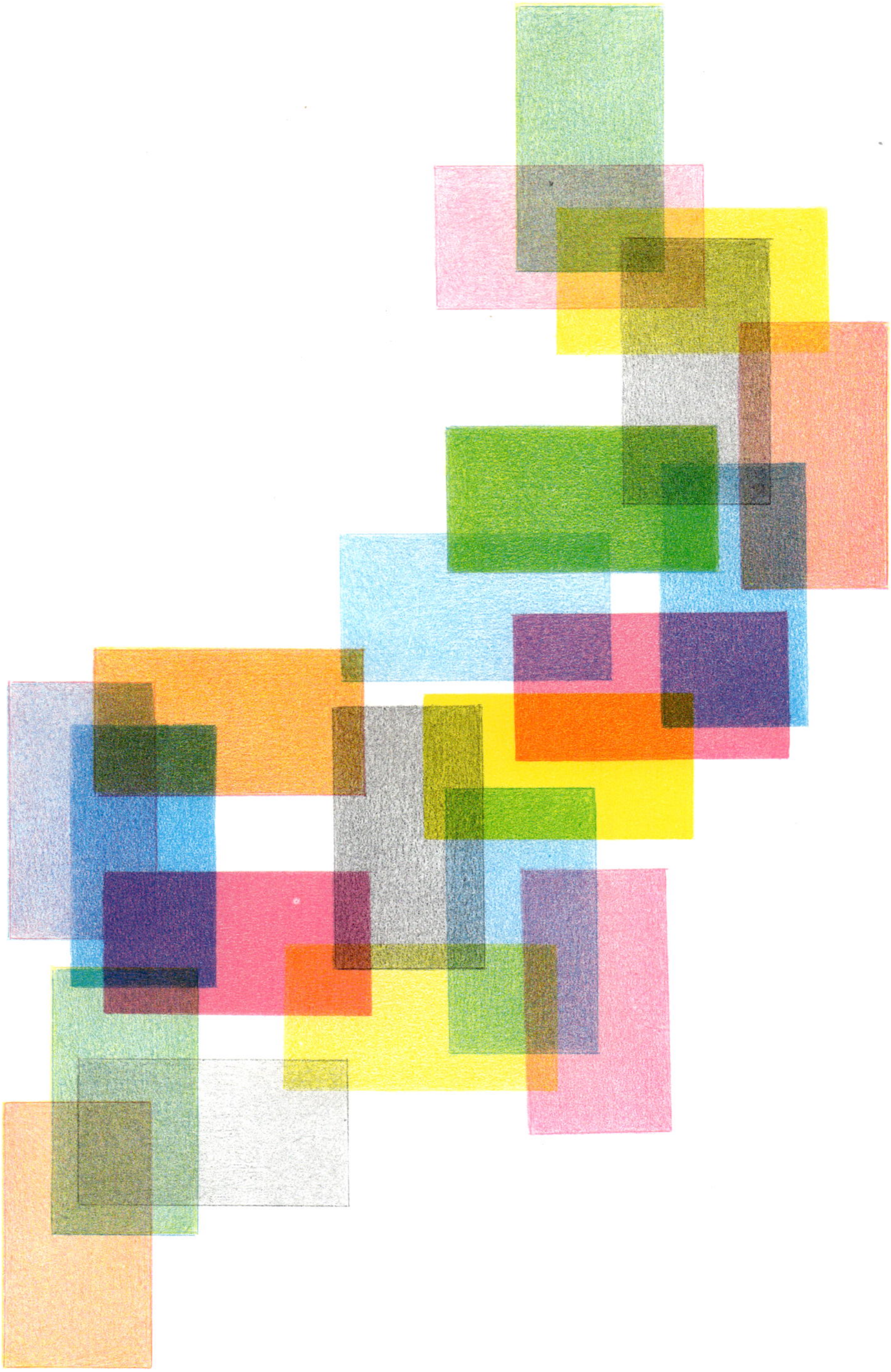

COLOR
IN REPRODUCTION

Theory and Techniques for Artists and Designers

HILDA SIMON

A Studio Book
The Viking Press · New York

First published in 1980 by A Studio Book/The Viking Press
625 Madison Avenue, New York, N.Y. 10022

Published simultaneously in Canada by
Penguin Books Canada Limited

Library of Congress Cataloging in Publication Data
Simon, Hilda.
Color in reproduction.
(A Studio book)
1. Color-printing. 2. Color in art. I. Title.
Z258.S58 686.2′3 79-20672
ISBN 0-670-22999-7

Printed in the United States of America by
Rae Publishing Co., Inc., Cedar Grove, N.J.
Set in Linotype Primer

Page 10 constitutes an extension of the copyright page

Contents

The Fundamentals of Color	13
The Basic Black	29
Black Plus One Color	41
Black Plus Two Colors	49
Four Colors for Full Color	57
The Infinite Variety: Color Combinations	73
The Art of Separation	89
Adventures and Challenges	103
Epilogue	121
Indexes	124

List of Illustrations

Light primary and secondary relationship 12
Visible spectrum 16
Separation of white light by prisms 17
Light primaries 18
Pigment primaries 19
Diagram demonstrating additive nature of light primaries 20
Diagram demonstrating subtractive nature of pigment primaries 21
Selective reflection of pigments in colored light 22–23
Color wheel of pigment primaries and secondaries 24
Reflection and transmission of colors 25
Woodcut by Albrecht Dürer 28
Copperplate engraving by Albrecht Dürer 31
Wood engraving by Thomas Bewick 32
Photographic halftone and wood engraving compared 33
Line pencil drawing 34
Line pen-and-ink drawing 34
Wash halftone illustration 35
Pencil halftone illustration 35
Black photographic halftone 36
Line illustrations printed in neutral black 38
Same illustrations printed in toned black 39
Two-color preseparated drawing in green and black 40
Photomechanical duotone in brown and black 42
Black halftone printed over brown screen tint 43
Black halftone marked for percentages of second color 44
Same halftone with second color printing as marked 45
Preseparated two-color drawing in black and brown 46
Preseparated two-color drawing in black and blue 47
Three-color illustration in magenta, blue, and black 48
Black halftone with hand-rendered two-color overlays 50–51
Black illustration marked for percentages in two colors 52
Same illustration with colors printing as marked 52
Illustration from p. 52 printing with softer colors 53
Preseparated three-color screenless drawing 54
Full-color process halftone of metal ornaments 56
Six-color preseparated screenless drawing 58
Full-color process halftone of a flower 60
Full-color process halftone of an interior scene 61

Detail of rose shown on p. 60 62
Screen angles for process separations 62
Photographic filters and the resulting process separations 63
Process reproduction of an oil painting by Georges Seurat 64
Full-color process halftone of a butterfly on a flower 66
Same butterfly in four-color screenless preseparation 67
Preseparated full-color illustration of a peacock 68–69
Four-color combination design 72
Classic example of color illusion 74
Enlargements of three halftone screen values 75
Optical color illusion: identical but different-looking hues 76
Optical color illusion: different but identical-looking hues 77
Example of clashing colors 78–79
Compartmentalization, or stained-glass effect 80
Color combination charts 82–87
Full-color screenless preseparation 88
Model for preseparated screenless illustration 90
Black separation 93
Yellow separation printed in yellow and also in black 95
Red separation printed in red and also in black 97
Blue separation printed in blue and also in black 99
Combination prints of yellow, red, and blue separations 100
Combination print of all four separations 101
Bird with plumage containing subtle hues of all colors 102
Brilliantly colored tropical bug 104
Shells with polished-looking colors 105
Delicate-hued orchid 107
Strongly-colored tiger orchid 108
Orchid blossoms with wide range of color combinations 109
Early American brass powder horn 110
Early American glass perfume bottles 110
Antique pewter-and-ceramic beer mug 111
Example of preseparation with fine lines printing on dark background 112
Brown resulting from superimposing same image in three colors 114–15
Deep purple resulting from use of same image for red and blue 116
Slate color achieved by superimposing same image in three colors 117
Iridescent tropical butterfly in screenless illustration 118
Preseparated drawing of three iridescent beetles 119
Screenless preseparation of hand-carved rococo angel 120
Preseparated illustration of iridescent sunbird 123

Acknowledgments

More than perhaps most others, this book has demanded a large measure of cooperation from the persons who work "behind the scenes" on preparations for any publication. Although primarily based upon knowledge and experience accumulated by me over the course of many years, the text had to be augmented by many items of information gathered from various segments of the graphic-arts field. Both in my attempts to devise an effective and attractive physical format for the book, and in my efforts to present the material as concisely as possible and in the most lucid sequence, I received generous assistance from those entrusted with the publication of the book. Because I greatly appreciate these contributions, I wish to express my appreciation not only in general terms, but also in the form of specific tributes to the persons most directly involved.

First in line is Barbara Burn, Editorial Director of Viking Studio Books. Her initial interest in my plans for this venture led her to select it as a prospective Studio Book, and she remained interested in every phase of the work on the manuscript and illustrations even as she became immersed in the countless details of coordinating the necessary editorial, production, and publicity efforts. Similarly, I wish to thank the others at Studio Books: Gage Cogswell, the Managing Editor, who handled the manuscript; Michael Shroyer, the Art Director, who was in charge of type selection and design; and the Production Manager Peter Grant, whose courteous cooperation and meticulous attention to production details were of great help to me.

A special measure of appreciation must go to Albert Margolies, Director of Production for The Viking Press. His genuine enthusiasm for the project was gratifying, and the excellent suggestions he made for the inclusion and clarification of various points of information added to the book's scope and interest.

Last but certainly not least, my thanks go to the people of the Rae Publishing Co., Inc., Lithographers of Cedar Grove, New Jersey, the lithographers entrusted with the printing of this book. Most especially, I wish to thank Edward R. Bathmann, Vice President of the firm, for his unfailing cooperation and his expert advice on a number of color-reproduction questions pertaining to both the text and the selection of specific pictorial examples. From the very beginning, when my own concept of the book's final form was still incomplete, Mr. Bathmann expressed not only interest but great confidence in the venture; throughout the subsequent months of work on text and illustrations, he was at all times willing to help in every possible way.

The many different and diverse facets of this book have been successfully combined into what I believe to be a well-balanced and useful as well as attractive sequence of information about printed color reproduction. In no small measure, this success is due to the contributions of the persons named above.

I also would like to express my appreciation, for permission to use in this publication illustrations from the books listed below, to the following publishing firms:

Dodd, Mead & Company for the illustrations on pp. 50, 51, 54 from *The Date Palm, Bread of the Desert* by Hilda Simon (pp. 40, 56, 98), Copyright © Hilda Simon, 1978; on pp. 68–69, 102 from *The Courtship of Birds* by Hilda Simon (pp. 66–67, 78), Copyright © Hilda Simon, 1977; and on pp. 118, 119, 123 from *The Splendor of Iridescence* by Hilda Simon (pp. 198, 221–23), Copyright © Hilda Simon, 1971. Used by permission of Dodd, Mead & Company.

Hart Publishing Company for the illustrations on pp. 110, 111, 120 from *The Joys of Hunting Antiques* by Stefan Salter (pp. 20, 58, 78, 183); Copyright © Hart Publishing Company, 1971. Used by permission of Hart Publishing Company.

J. B. Lippincott Co. and Harper & Row, Publishers Inc. for the illustrations on pp. 107, 108, 109 from *The Private Lives of Orchids* by Hilda Simon (pp. 52, 70, 86), Copyright © J. B. Lippincott Co., Harper & Row, Publishers Inc., 1975. Used by permission.

The Metropolitan Museum of Art, New York, Bequest of Samuel A. Lewisohn, for the illustration on p. 64, Georges Seurat: *A Sunday Afternoon at the Grande Jatte.* All rights reserved, The Metropolitan Museum of Art.

The Vanguard Press for the illustrations on p. 46, from *Exploring the World of Social Insects* by Hilda Simon (p. 52); Copyright © Hilda Simon, 1962; on pp. 88, 92 from *Milkweed Butterflies* by Hilda Simon (pp. 56, 60), Copyright © Hilda Simon, 1962; on p. 105 from *Snails of Land and Sea* by Hilda Simon (pp. 58, 84, 115), Copyright © Hilda Simon, 1967. Used by permission of Vanguard Press.

The Viking Press for the illustrations on pp. 32, 33 from *Wood Engraving* by David M. Sander (pp. 16, 20), Copyright © David M. Sander, 1978; on p. 34 from *Archosauria* by John C. McLoughlin (p. 40), Copyright © John C. McLoughlin, 1979; on p. 35 from *Figures/Faces* by Hugh Laidman (p. 142), Copyright © Hugh Laidman, 1979; on p. 35 from *Alfred R. Waud: Civil War Artist* by Frederic E. Ray (p. 181); on p. 48 from *Partners, Guests, and Parasites* by Hilda Simon (p. 78), Copyright © Hilda Simon, 1970; on p. 104 from *Insect Masquerades* by Hilda Simon (p. 78), Copyright © Hilda Simon, 1968; on p. 112 from *Dragonflies* by Hilda Simon (jacket illustration), Copyright © Hilda Simon, 1972; on pp. 114, 115, 116 from *Snakes: The Facts and the Folklore* by Hilda Simon (pp. 12, 14, 18), copyright © Hilda Simon, 1973; on pp. 114, 117 from *Feathers Plain and Fancy* by Hilda Simon (pp. 12, 35), Copyright © Hilda Simon, 1969. Used by permission of Viking Penguin Inc. On pp. 56, 60, 61, 66 from *Eisenstaedt's Guide to Photography* by Alfred Eisenstaedt (pp. 58, 126, 154, 158), Copyright © Alfred Eisenstaedt, 1978. Used by permission of the author.

COLOR
IN REPRODUCTION

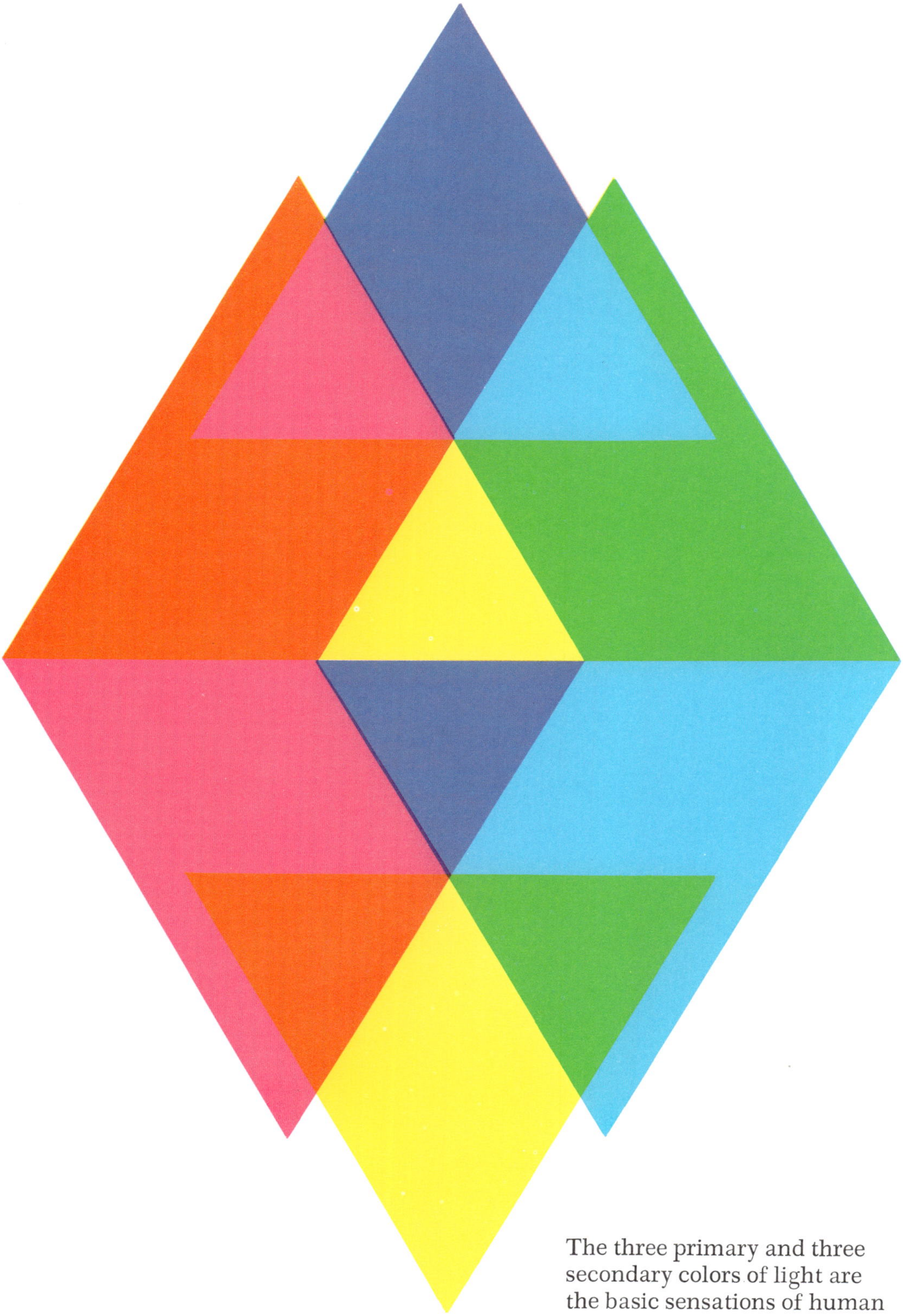

The three primary and three secondary colors of light are the basic sensations of human color perception.

The Fundamentals of Color

The elusive phenomenon of color seems to have fascinated human beings from the beginning. That various ancient civilizations attributed special—often mystical or religious—properties to certain colors is a matter of historical record. Some early societies chose crimson, the color of blood and by extension of courage and of life itself, as their royal color in preference to the famous "imperial purple." To this day, the "red carpet" is a symbol of privilege and preferential treatment.

Even in the mystery-shrouded times known as prehistoric because no written record exists of them, *Homo sapiens* appears to have appreciated and used color to enhance various aspects of everyday life. Although man shares a number of physiological features with the other primates, his pronounced color sense is one item on a long list of particulars in which he differs decisively from them. This is especially remarkable because the anthropoid apes, unlike most other mammals, apparently possess the same range and scope of color vision as human beings. Yet observations of their behavior in the wild would seem to indicate that primates such as gorillas and chimpanzees have very little, if any, native feeling for color in their normal daily life. This inherent lack among the great apes is in striking contrast to the color sense found in many lower-ranking animals such as birds and insects, some of which are influenced and guided by color in such vital activities as selecting mates or finding food. The strongly developed color appreciation of the fabulous bowerbirds of Australia, and the keen discrimination they display in collecting suitably colored ornaments for their elaborate courtship bowers, is only one case in point.

As for mankind, we find that the artistic endeavors of prehistoric humans offer the most enlightening clues to the personality—as opposed to the physiology—of people about whom we otherwise know next to nothing. Artifacts and paintings such as those from the Cro-Magnon caves of France show that these ancients had a remarkable eye for form and color, the artistic ability to transfer their view of three-dimensional objects to two-dimensional surfaces, and the ingenuity to utilize substances from their natural environment as colors for their paintings.

As civilizations began to develop, so did the creative urge of man to reproduce in handmade images not only the objects and creatures of his natural environment but those of his imagination, dreams, and religious aspirations as well. Three-dimensional reproduction was realized in architecture, sculpture, and related crafts; two-dimensional reproduction dealing with outline and, especially, color was effected in drawings, paintings, and such crafts as weaving and embroidery.

Until fairly recently any artist also had to be a craftsman, and the arts and crafts were so closely linked that often there was little distinction between them. Many of the greatest painters and sculptors not only started out as artisans but remained practicing craftsmen all their lives. Michelangelo began his career as a stonemason, and the famed German painter Albrecht Dürer as an apprentice goldsmith. In later life Dürer mastered skills in a variety of other crafts. Far from being ashamed of their "workman" status, these artists were proud of it, frequently rating it more highly than their artistic talent, which many of them took more or less for granted. Constantly laboring to broaden their technical, mechanical skills so they could employ a variety of means to serve their artistic goals, these men were never content to accept existing materials and techniques but always attempted to improve upon them. For painters, this meant primarily a quest for new color effects, for which they sought out purer coloring substances that would prove stable and resist fading. Artists such as Leonardo da Vinci spent a great deal of time and energy experimenting with new pigments, mixing and testing special materials, not only because they believed these pigments would give finer effects but also because they hoped they would resist fading and retain their brilliance over the years. The latter goal was a great concern to painters in the past, especially because they did not know why certain compounds were stable and "colorfast," whereas others, pure and glowing when applied, lost much of their strength with the passage of time and became dull and weak.

The frustration resulting from the problem of fading is easy to appreciate when we keep in mind that the artists of those days did not know what color was. Although they had extensive experience with the mixing and handling of coloring substances, and could employ color with superb skill in their paintings, they did not understand color as a purely optical phenomenon. Until 1666 color was believed to be an integral part of the object displaying it. In that year Sir Isaac Newton discovered that white light is made up of a band of different colors and proved it to skeptics by demonstrating how a beam of light could be separated into its component colors with the help of a prism. Newton thereby established that color is nothing but light, and that objects are colored only because they reflect some portion of the color band, or spectrum, that makes up white light. This discovery invested the phenomenon of color with a new scientific importance rather than just an aesthetic one, and physicists since Newton have never ceased to study its many aspects. In a parallel development the properties of pigments—substances capable of producing color by selectively reflecting just one portion of white light—also were subjected to continuing scientific scrutiny.

Despite the overwhelming amount of information on the twin concepts of light and color accumulated by physicists since Newton's time, much about these elusive quantities remains a mystery even today. One may be aware of the fact that Newton's corpuscular theory of light was replaced by the wave theory of the Dutch physicist Huygens, and that the latter theory was at least partly superseded by Max Planck's quantum theory early in this century. One may also be vaguely familiar with the fact that quantum physics is at the core of atomic science without in the least understanding the role of light and color in that particular context.

A good many basic facts about these phenomena have become

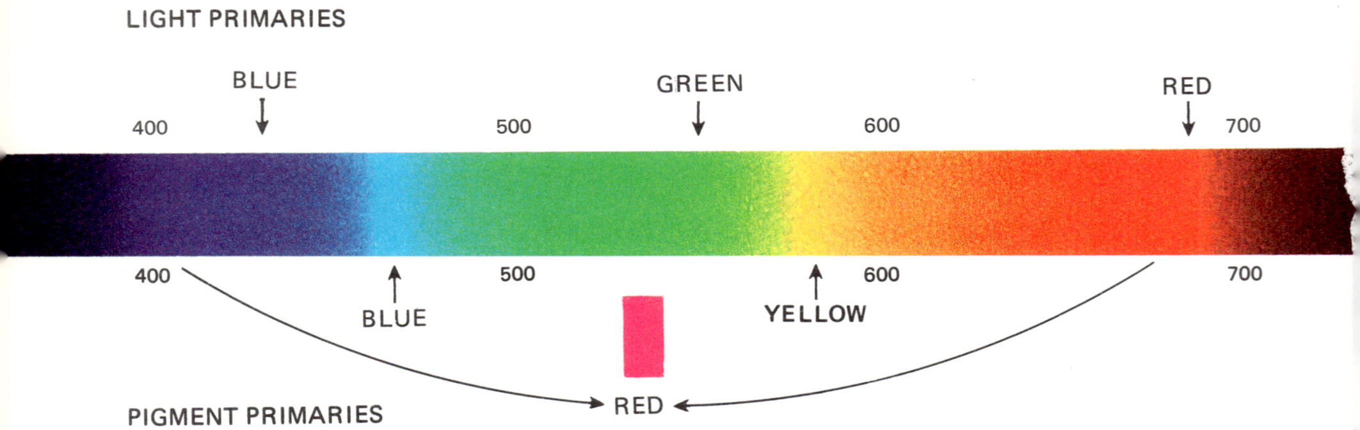

Bordered by invisible infrared and ultraviolet, the part of the spectrum visible to the human eye ranges from long-wave red at one end to short-wave violet at the other.

common knowledge, however. Everyone is familiar with what we call the visible spectrum, which ranges from the long-wave red at one end to the short-wave violet at the other. We also know that this spectrum is but a tiny portion of the overall spectrum of radiant energy, most of which we cannot perceive. The prefixes *infra* and *ultra* denote colors beyond the range of our vision but not necessarily beyond that of other creatures. Thus ultraviolet is a visible color to bees, some butterflies, and certain shellfish such as crabs.

The main sensations of human color vision—about whose mechanics we still know very little—are defined in terms of *primaries*. The *fundamental,* or psychological, primaries are thought to be red, yellow, green, blue, black, and white. Three of those six—red, yellow, and blue—are designated the *pigment* primaries of painting and color reproduction because all other hues may be mixed, at least theoretically, from those three colors. Black and white are also important in visual reproduction for light and shadow definition.

Like pigment primaries, the primaries of *light* also number three; and are usually listed as red, green, and blue. On the face of it, two of the light primaries—red and blue—appear identical with two of the pigment primaries. That, however, is an erroneous impression resulting from imprecise terminology. Using the broad terms "red" and "blue" without qualification for both sets of primaries often causes confusion and a failure to appreciate the special relationship between light primary colors, which are *source* light, and pigment primary colors, which result from *reflected* light.

Since practically our entire color experience is based upon reflected light—the human eye is poorly equipped for looking directly at, or into, sources of radiant energy—most people are much more familiar with pigment colors and their qualities than with the components of the spectrum. The physicists' statement about "balanced sets of red, green, and blue" combining to make white light has become something of a cliché and is often repeated without any real notion of what it involves. What *kinds* of red, green, and blue combine magically into the colorless light we call white when mixed in the correct proportions? The specific grounds upon which the Young-Helmholtz trichromatic theory of light is based are generally not understood, yet the basic premises of the theory are well worth a closer look to all those concerned with color reproduction.

The separation of white light into its component colors, or wavelengths, with the help of prisms offers proof of its composite nature.

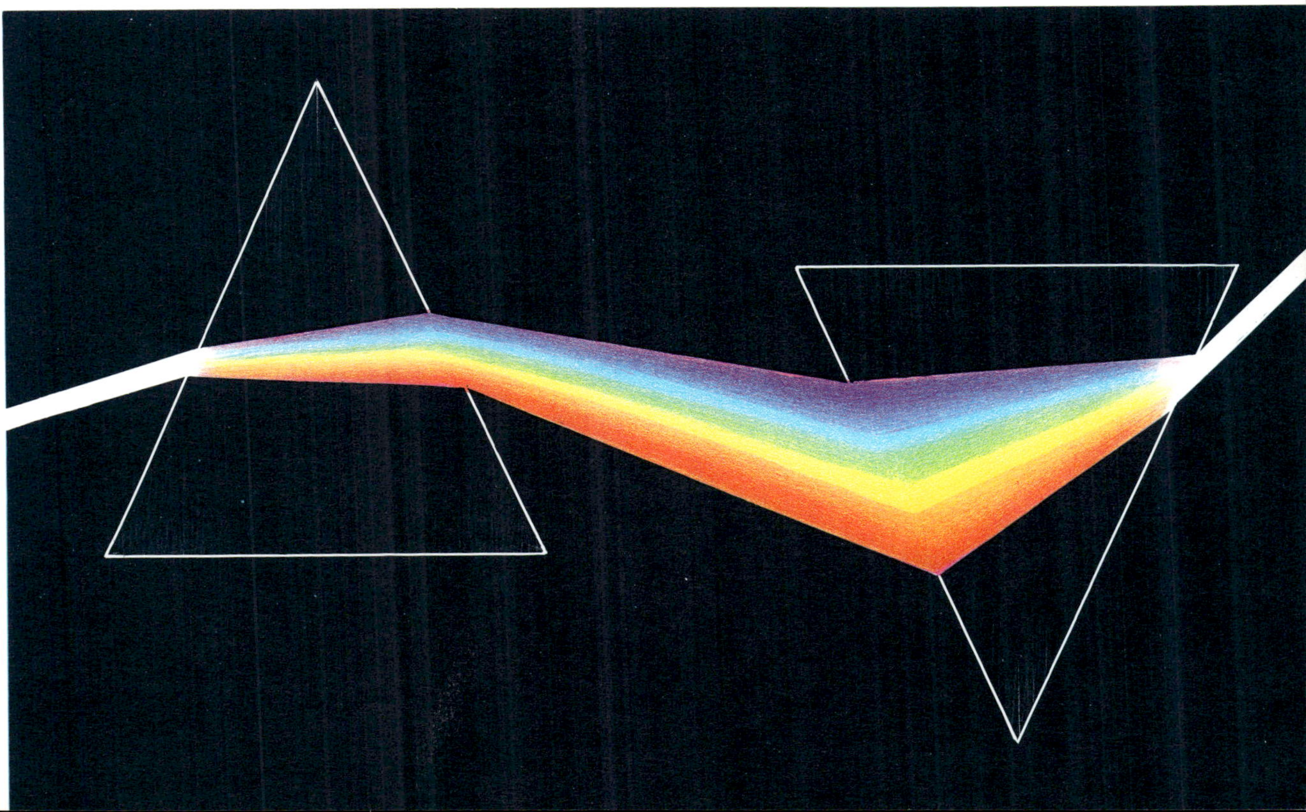

The colors of the spectrum represent light of different wavelengths and are measured by millimicrons, whose symbol is $m\mu$. The spectral color scale ranges from approximately 780 $m\mu$ at the long-wave end to 380 $m\mu$ at the short-wave end; beyond are the invisible rays of, respectively, infrared and ultraviolet. However, toward the extreme ends of the visible spectrum there occurs an increasing admixture of these invisible rays and a corresponding decrease in color intensity. For practical purposes, then, it may be stated that the spectrum's color range extends from about 400 $m\mu$ to 740 $m\mu$.

Physicists pinpointed the three light primary colors after experiments showed that balanced amounts of red, green, and violet-blue yielded white light. Optical identification points for these colors were set at 691 $m\mu$ for red, 546 $m\mu$ for green, and 437 $m\mu$ for blue. Seen on a reproduction of the spectral color scale, these colors may be described as a deep scarlet red, a yellowish grass green, and a deep violet-blue. The spectrum's other colors are varying mixtures of any two of these primaries—orange, for example, results from large amounts of red plus small amounts of green. *Secondary* colors appear wherever the mixture of two primaries is well balanced; yellow is such a red/green combination, and cold "true" blue a similar mixture of green and violet-blue. The third secondary, a mixture of red and violet-blue commonly known as

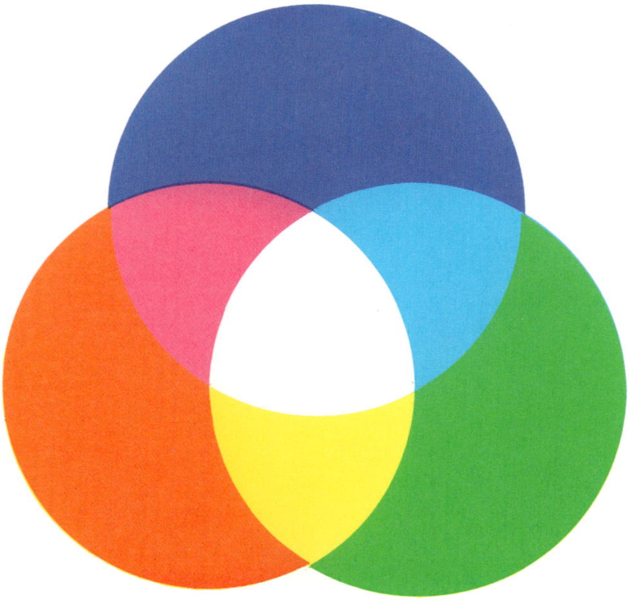

The three primaries of light. Each set of two produces a secondary color; all three combined yield white light.

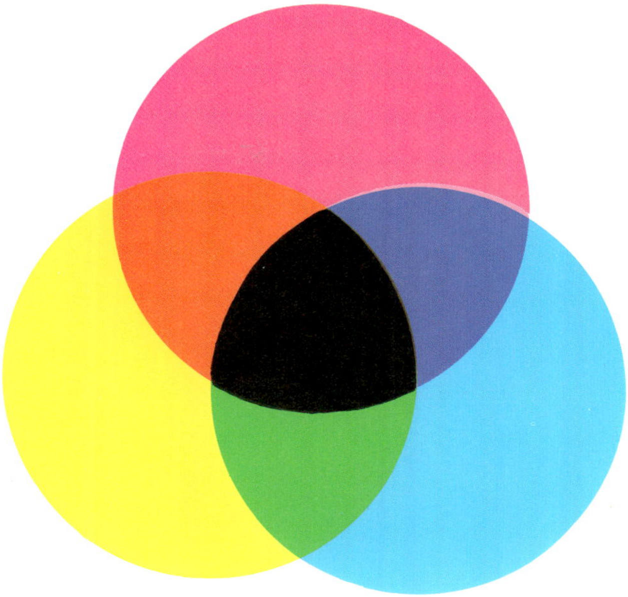

The three pigment primaries. Each set of two produces a secondary color similar to a light primary; the combination of all three results in black.

magenta, does not show on the spectral color scale because its component colors do not adjoin but are located at opposite ends of the spectrum.

According to the trichromatic light theory, each of the secondary colors contains balanced amounts of two primaries and needs only the third primary for the full complement of white light. Experiments confirm that theory; balanced amounts of red and green produce yellow light, and its addition to a beam of violet-blue yields white light. Adding the appropriate secondary, or *complementary,* color to the other primaries gets the same result.

Perhaps the most obvious visual feature of the light complementaries is their resemblance to the pigment primaries, and the fact that mixtures of any two of the latter on the palette or in the watercolor box yield hues that correspond to the colors of the light primaries. Most people know from their earliest experience with color mixing that yellow and magenta make bright scarlet red, whereas yellow and blue produce green, and magenta and blue a violet-blue color. This clearly proves that the colors of light and those of pigments are exact opposites, the unmixed hues of one being the mixed hues of the other, and vice versa. Consequently,

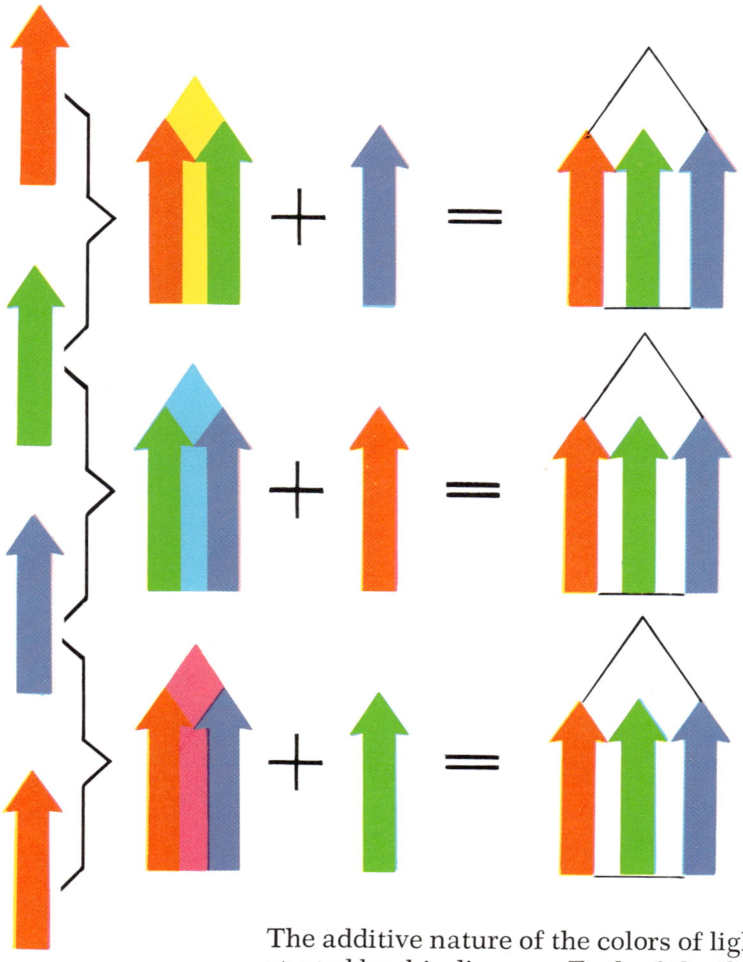

The additive nature of the colors of light is demonstrated by this diagram. Each of the three light primaries on the left consist of a single color; combining any two of them creates secondaries, each of which contains two colors. By adding the third primary, balanced sets of red, green, and blue are formed; they combine to make white light.

the combination of each set of primaries also produces the opposite effect: that of light results in white, or total, light; that of pigments in black, or total elimination of light. This principle (often poorly understood) is the basis of all color reproduction by pigments, regardless of whether such reproduction is achieved individually by hand, as in a painting, or mechanically by machines geared to mass color reproduction.

The fundamental difference between the two sets of primaries is that of source versus reflected light. The primaries of light are *additive* because the addition of another color to a beam of colored light results in more light, with all three primaries together producing the full complement of white light.

By the same token, pigment primaries are *subtractive* because any addition of other colors results in a reduction of the amount of light reflected by the combination. A pigment can produce a color only by selectively reflecting one portion of the spectrum while absorbing the rest. The primaries magenta, yellow, and blue each reflect two-thirds of white light, meaning two light primaries. The third is absorbed and so subtracted from the total. This works out as follows:

Magenta reflects Red plus Violet-blue but subtracts Green.
Yellow reflects Red plus Green but subtracts Violet-blue.
Blue reflects Green plus Violet-blue but subtracts Red.

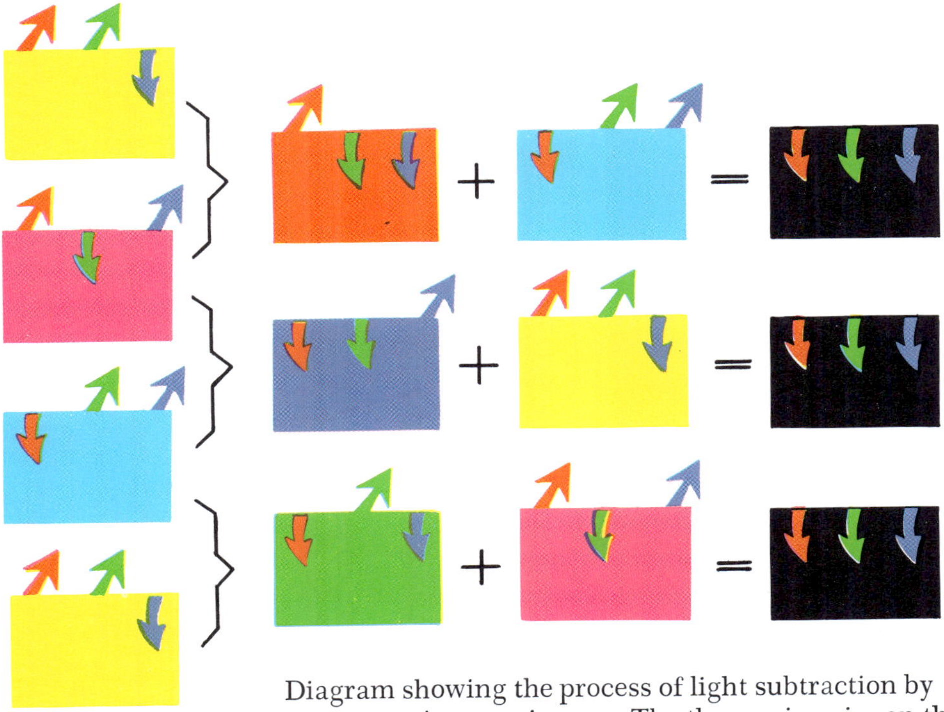

Diagram showing the process of light subtraction by pigment primary mixtures. The three primaries on the left reflect two colors and absorb one each. Combining any two primaries yields secondary colors that reflect only a single color each, while absorbing two. Adding the complementary primary to any secondary leads to the absorption of the third color, which leaves no light to be reflected.

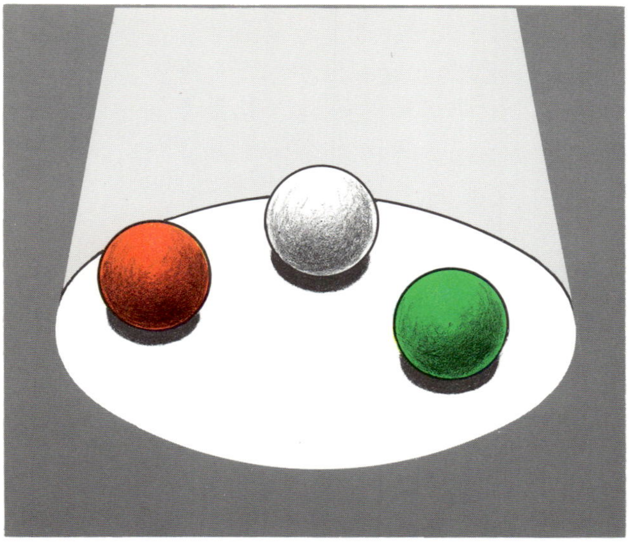

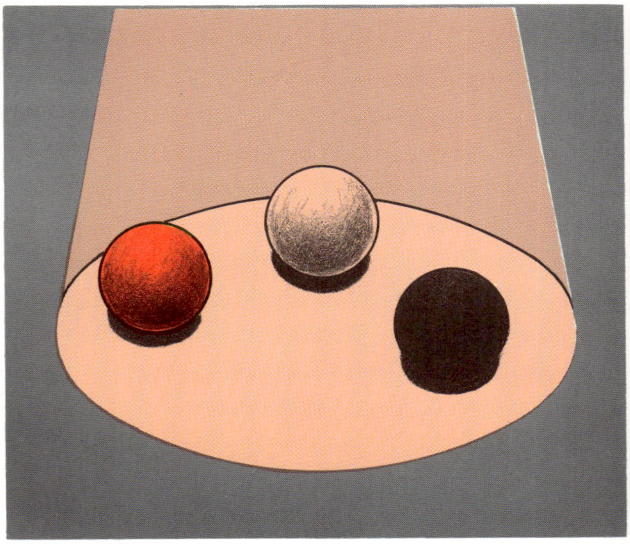

If any two pigment primaries such as yellow and blue are combined, a second light primary is subtracted from the amount of reflected light, for each absorbs one of the other's colors, yellow canceling out the violet-blue content of blue, and blue eliminating the red in yellow. It follows that green is the only light primary left to be reflected. The same principle holds for the combination of magenta and blue, in which red and green are alternately subtracted to leave only violet-blue, and of magenta and yellow, which eliminate the green and violet-blue to produce scarlet red. This means that any hue resulting from a balanced mixture of two pigment primaries can reflect only one-third of the white light that illuminates it. Actually, the one color present in both pigment primaries becomes visible only after the other component colors are removed by absorption in the mixture.

Although this may sound complicated, it is quite simple, as a look at the accompanying diagrams of light and color mixtures will show. For many people, the main difficulty lies in the fact that more means less in pigment color mixtures because each additional color reduces the amount of reflected light. The more the various pigment colors are mixed, the less bright and intense the resulting hue will be. Even before the underlying physical principle became

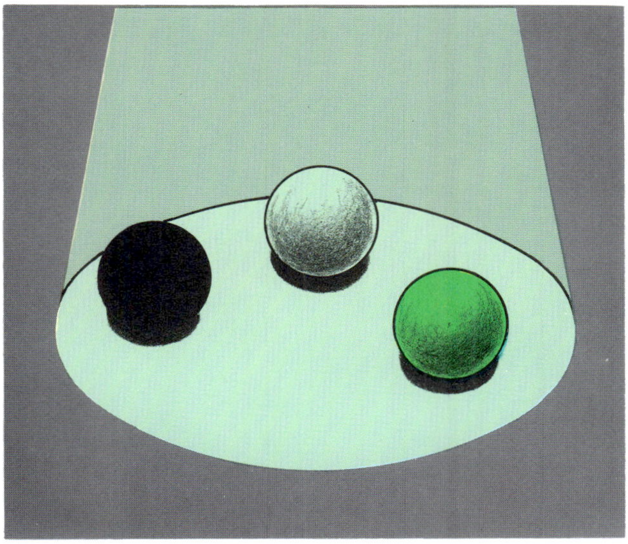

Because any object can reflect only its own color, white light is needed to render visible the entire range of the spectrum. In this example, each beam of colored light is reflected only by the object whose hue corresponds to the beam's wave-lengths.

known, all great painters were acutely aware of this fact and used it with consummate mastery in their paintings.

Anyone interested in confirming by experiment the premise that mixed pigment colors such as green or "warm" red reflect only one-third each of white light may do so quite easily. Place a red and a green object on a white surface and then illuminate them, first by a beam of white light, and then, by individual beams of red and green light successively. Each object will display its own hue under the white light. But in the colored light, only the object whose color corresponds to that of the beam will appear colored; the other will look black or dark gray. The white surface, on the other hand, will assume the hue of the illuminating light in each instance because white reflects any and every color of the spectrum. This experiment proves not merely that an object reflects only the color or colors of its pigments but also that pigments by their nature are subtractive of light.

The next logical step is the combination of all three pigment primaries. If the pigments are used in full strength, all three light primaries will be eliminated and thus there will be no light left to be reflected. The result is the complete absorption of all light waves commonly referred to as the "color" black. Or at least that is what happens in theory; in practice, the impurities in the pigments usually produce a certain amount of reflection "leakage" so that the resulting hue is a deep blackish brown rather than a perfect black. In that respect, ink pigments used in printing are notoriously defective, sometimes alternately reflecting or absorbing more light than they should.

Once the subtractive principle of all pigment color combina-

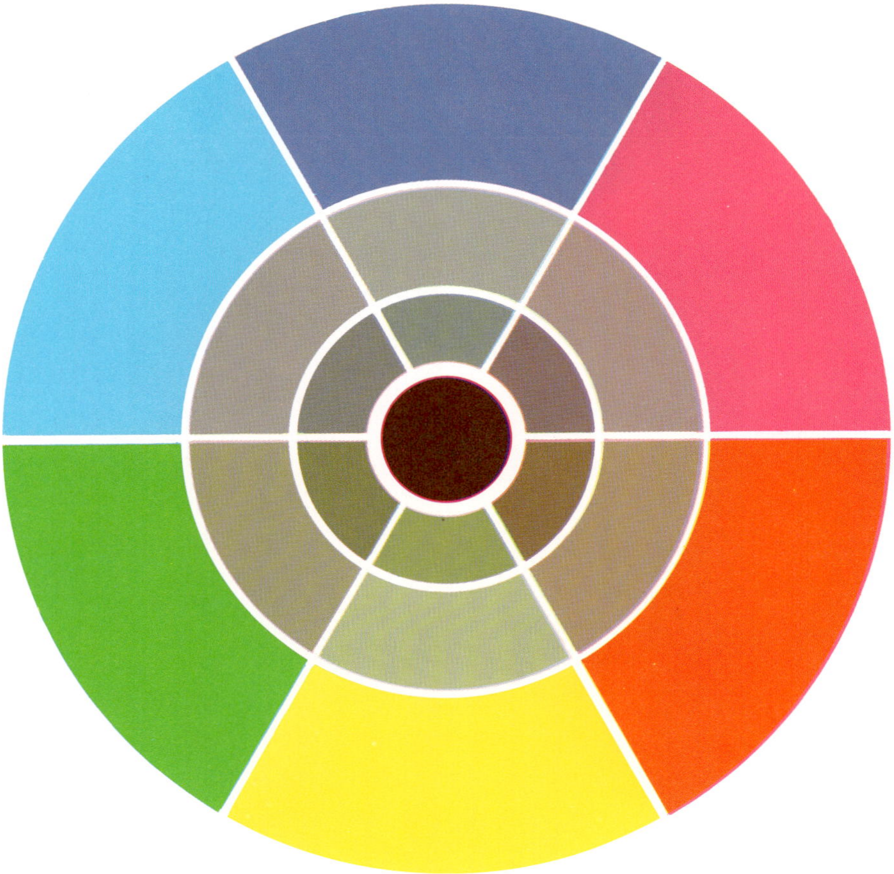

This color wheel shows each of the three primary pigment colors opposite its secondary, or complementary, color. The inner rings represent combinations of each primary-secondary set in increasing percentages. The full-strength mixture of all three colors in the center yields black.

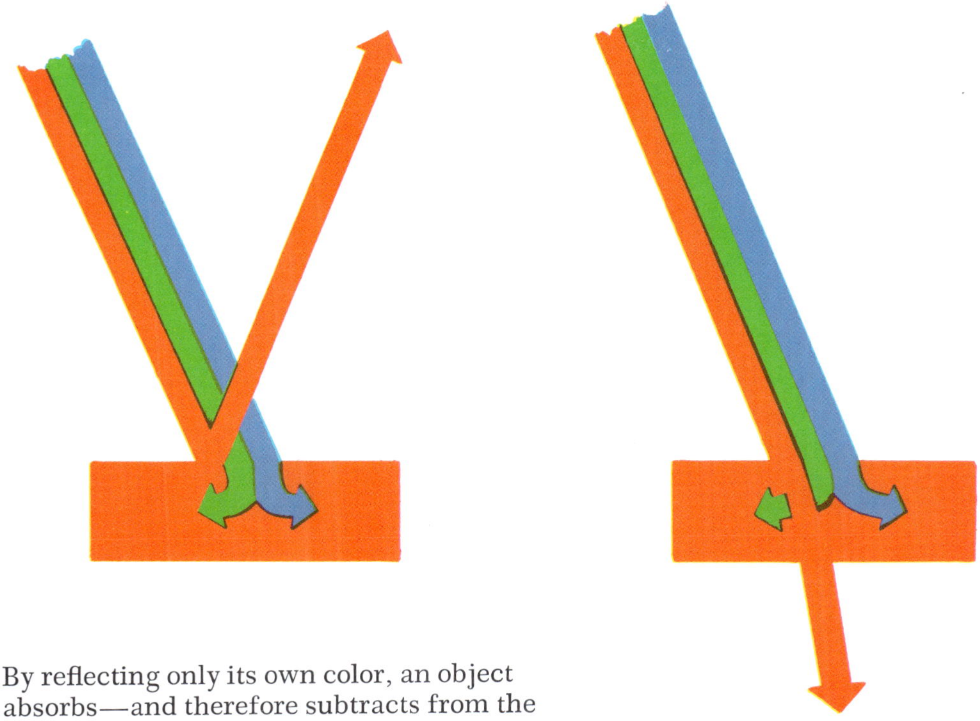

By reflecting only its own color, an object absorbs—and therefore subtracts from the total amount of reflected light—all other wavelengths. Similarly, a filter transmits no light except that portion corresponding to its own color.

tions has been grasped, many frustrating and puzzling aspects of color reproduction can be resolved. The understanding of just what a color does in any given combination provides artists with the knowledge they need to deepen, subdue, or otherwise manipulate their colors. Many too-bright or harsh colors can be softened by adding precisely calculated percentages of other hues designed to subtract some of the unwanted reflected light ingredients and create more attractive shades.

Selective reflection is not the only means by which portions of white light may be eliminated; filters that *transmit* rather than reflect light do the same. Just as pigments reflect only the light waves of their own color, so filters transmit only the waves of theirs: a yellow filter absorbs all except the yellow light waves. The filter eliminates the blue-violet portion of the spectrum; the remaining red and green light waves combine to form the mixture yellow. By the same token, red and green filters eliminate two-thirds each of white light and permit only the waves of a single primary, or color, to pass through. The filter principle is seen daily in our traffic lights; it also happens to be the first and fundamental step in the process of photomechanical color reproduction.

In addition to pigment colors, there are those hues known as structural colors because they result from the refraction of light by submicroscopic surface structures occurring in both organic and inorganic matter. The colors of the rainbow, the "fire" of gemstones, and the shimmering hues of soap bubbles are only a few examples of inorganic iridescence. In living organisms structural colors are found primarily in birds and insects. Such "flying jewels" as hummingbirds, certain tropical butterflies, and iridescent metallic beetles are justly famed for their pure, glittering, shifting hues. There are, however, some structural colors that do not display such "rainbow" characteristics. The blue of the sky and that of most blue birds, to name only two examples, are noniridescent structural hues produced without the aid of any blue pigment.

Given today's advanced technological means, pigment colors and even noniridescent structural colors may be reproduced with considerable fidelity on paper. Iridescent colors are another matter entirely. These brilliant but elusive hues, unadulterated as they are by pigment impurities, cannot be satisfactorily recreated by printing processes. Entirely dependent on both the angle of light incidence and the angle of observation, they are too translucent, too much "pure light," and too changeable to lend themselves to reproduction, by pigment colors, on a flat surface such as paper. Creating the *illusion* of iridescence is a fascinating challenge for the artist and the printer alike.

It is possible, however, to produce good printed color images of most other objects in our environment. Depending on the quality of the original photograph or painting, on the materials used, and on the skills employed in the process, such printed reproductions range all the way from poor to excellent. To satisfy the considerable demand for good color illustrations, new ways of improving—or at least retaining—the quality of color reproduction without inflating printing costs are constantly being explored. Although the traditional black ink, for centuries the sole "color" in all printed texts and illustrations, remains the mainstay of printing, chromatic color in its various combinations has become an integral and vital part of our many-faceted world of books, magazines, and other literature.

EPITOME IN DIVAE PARTHENICES MARI
AE HISTORIAM AB ALBERTO DVRERO
NORICO PER FIGVRAS DIGES
TAM CVM VERSIBVS ANNE
XIS CHELIDONII

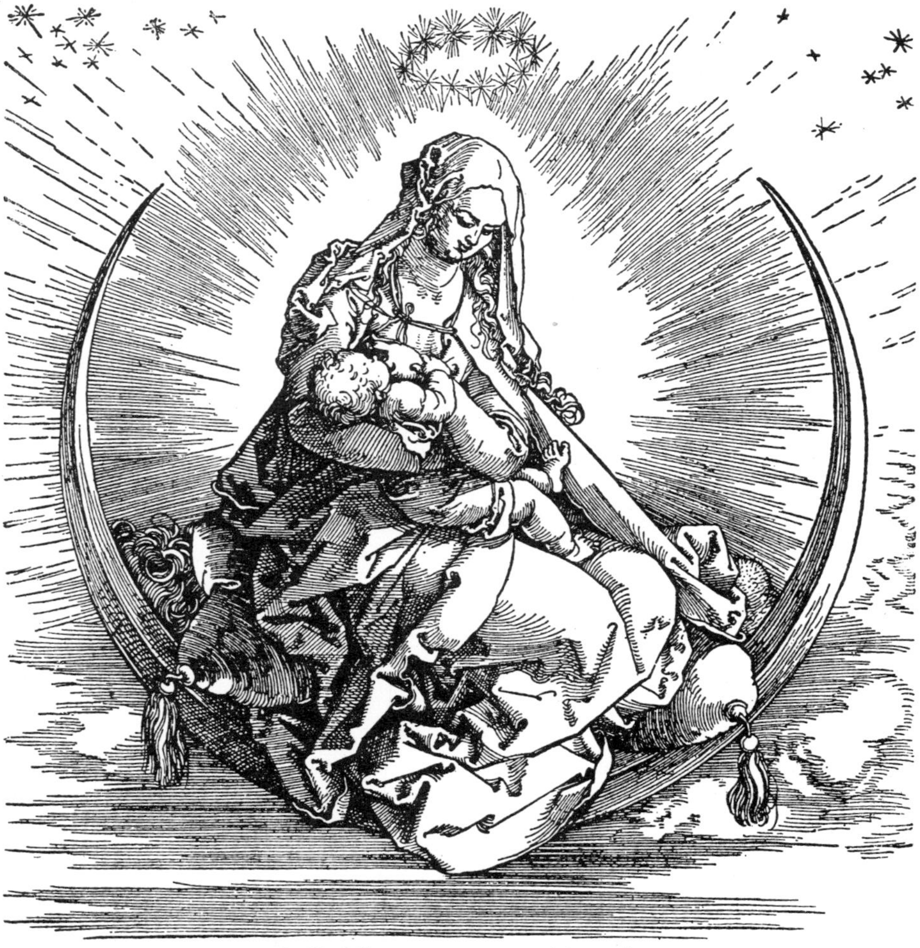

Quiſquis fortunæ correptus turbine.perfers
Quam tibi iacturam fata ſiniſtra ferunt.
Aut animæ delicta gemis.Phlegethontis & ignes
Anxius æternos corde tremente paues.
Quiſquis & vrgeris iam iam decedere vita
Alterius:migrans:neſcius hoſpitij.
Huc ades:auxilium:pete:continuoq; rogabo
Pro te:quem paui lacte:tuliq; ſinu.
Ille deus rerum mihi ſubdidit aſtra:deoſq;.
Flectitur ille meis O homo ſupplicijs.

TITELBILD ZUM MARIENLEBEN. 1511

Title page of Albrecht Dürer's book (1511) on the life of Mary. After making the woodcuts, Dürer had them reproduced by a printing firm in which he was a partner.

The Basic Black

For the great majority of printed reproductions, and certainly for book printing, black is the basic and often the only color. Any book can be printed in black without other colors, but it cannot be printed in other colors without black—or at least not without some color dark enough to substitute for black. Although there are exceptions—books employing large, heavy type, for example—for which a lighter color is used and black is omitted altogether, these are showcase pieces printed for special effects and usually in very limited editions, and therefore subject to conditions not applicable to the average book. By and large, the noncolor black is the most important "color" in printing, and the term "printers' ink" generally implies the black ink used for text matter today as it has been for the past five hundred years.

Printers' ink has a long and interesting history. Although its beginnings are shrouded in uncertainty, we may assume that it was developed from the writing inks used by the Egyptians and the Chinese four thousand years ago. The Chinese apparently were pioneers in perfecting the first printing inks.

The main ingredient in all old inks was soot or lampblack, which was mixed with certain oils, animal glues, and other substances. Many early printers had their own, jealously guarded "secret" ink formulas.

From the very beginning, good printed reproductions depended to a large extent on the quality of the ink employed in the process; it had to produce a clean, sharp impression and dry without smearing. Although the word "ink" is often associated with conventional writing fluids, printers' ink does not flow like a liquid;

instead, it is a heavy, viscous substance that must possess a precisely balanced combination of ingredients to print well and dry quickly.

By the time the German inventor Johannes Gutenberg perfected the printing press and movable type, which revolutionized book manufacture and initiated mass reproduction of written texts in the West from the middle of the fifteenth century onward, the Chinese had been making prints from type cast in metal for many years. The oldest such text still in existence was printed in Korea in A.D. 1397, almost half a century before Gutenberg introduced his epoch-making movable type in Europe.

Illustrations featured in books printed during the first century or so after Gutenberg's invention consisted almost exclusively of woodcuts, but copper and steel engravings became popular in the sixteenth century. In a woodcut the desired image is drawn on a carefully planed block of wood, and the spaces between the lines are tooled out with special utensils. The raised portions are then inked and pressed against paper for a black printed reproduction of the original drawing. The hollowed-out spaces stay free of ink and therefore remain white.

The opposite technique is used in copper and steel engravings. In the *intaglio,* or sunken, method of reproduction the tooled-out spaces represent the image; after the plate has been inked, all the ink on the raised portions is wiped off so that it remains only in the depressions below the surface. The print reproduced on paper corresponds to the intaglio image.

In the past many artists were masters in one or more of these engraving skills. Albrecht Dürer had his own workshop where he trained apprentices, and he was also a partner in a publishing firm that printed books illustrated entirely with his own woodcuts, for he considered book illustration a major art form. A century later Rembrandt acquired several of Dürer's woodcuts, which undoubtedly served as an inspiration to the great Dutch master for some of his own famous etchings.

In later years Dürer added both copperplate and steel engraving to the skills in which he excelled; a number of his best known smaller works are executed in those media. It reveals a great deal about the attitude prevailing in the Renaissance world that a painter as renowned in his own lifetime as Albrecht Dürer should consider these crafts as art forms fully equal, in their own way, to painting.

In those days engraving was the only means by which copies of famous paintings could be reproduced as printed images. The quality and accuracy of the engraved copy depended on the skill

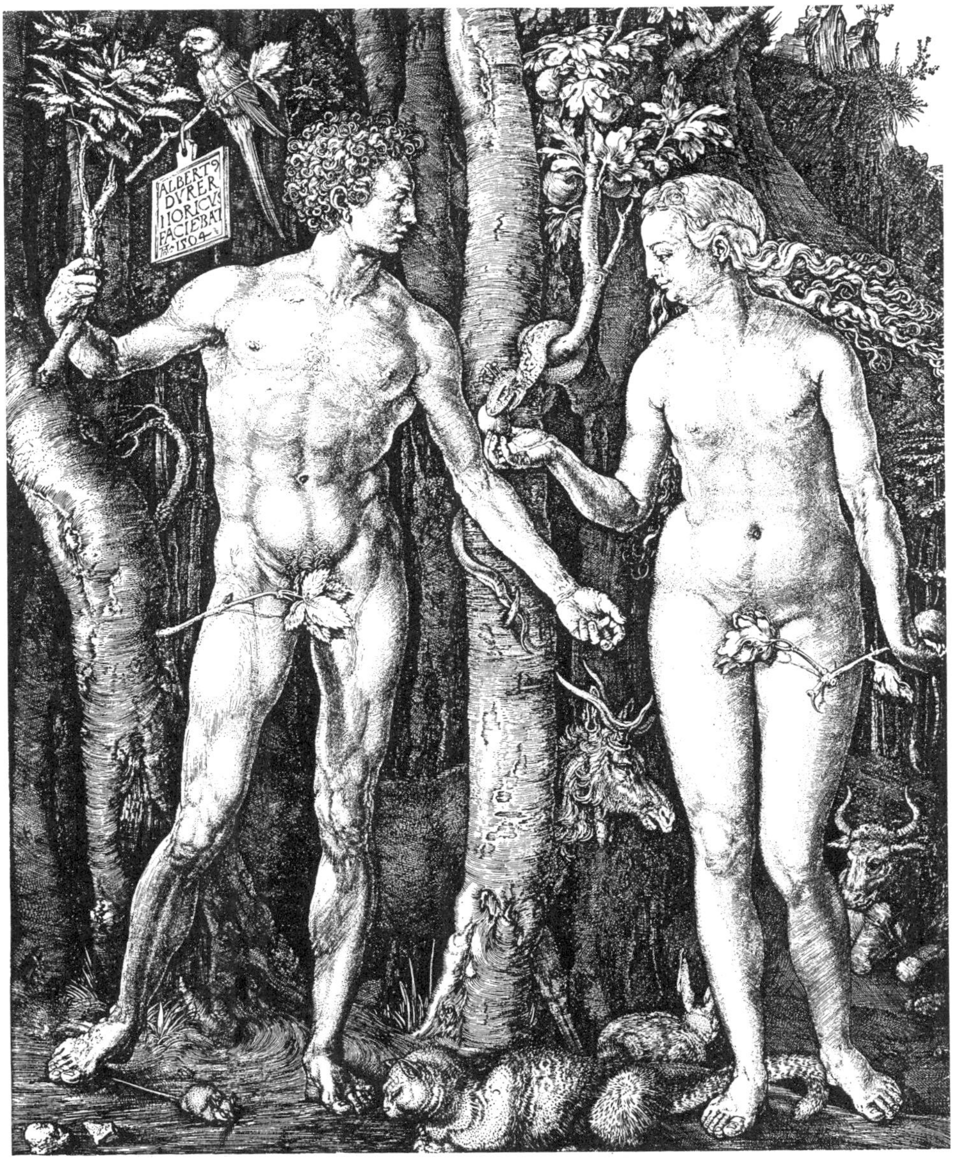

One of Dürer's famous copperplate engravings depicting Adam and Eve in the Garden of Eden.

and talent of the copyist, and many of them were quite good. There exist numerous woodcuts and steel engravings representing works by the Italian master Titian and other well-known artists. These printed reproductions permitted art lovers without access to the original at least an impression of what such a masterpiece looked like.

It was a black-and-white impression, of course, for practically no color was used in printing during the first few hundred years after Gutenberg's invention, although individual prints of engravings were sometimes hand-colored. Even a few books had colored illustrations that were produced in that way, with the artist usually hand-coloring the first few black-and-white imprints, and trained apprentices then copying his colors on subsequent sheets. This laborious process naturally limited colored editions because each volume was quite literally colored by hand.

Although copperplate and other engravings replaced woodcuts to a large degree after the end of the sixteenth century, woodcuts regained much of their popularity in the late eighteenth century. In 1770 the Englishman Thomas Bewick introduced a special technique for making very finely detailed woodcuts known as wood engravings, in which endgrain woodblocks were used instead of the traditional blocks cut with the grain. By tooling out fine, closely

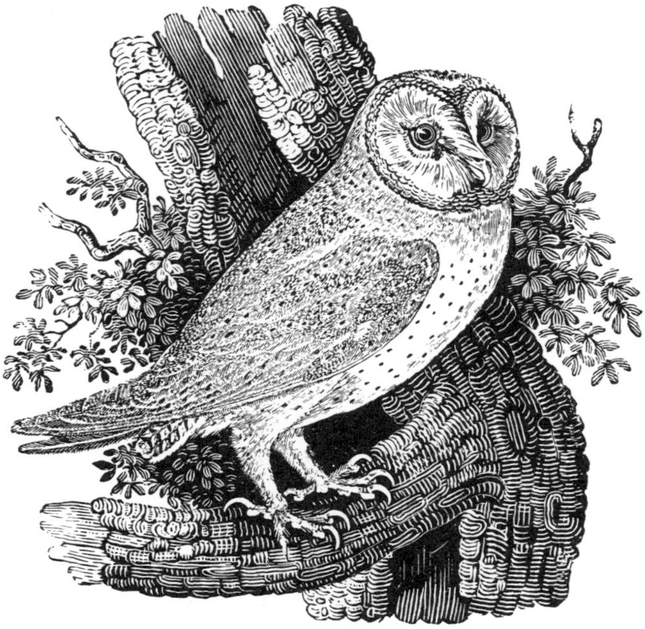

Wood engraving of an owl by the English artist-naturalist Thomas Bewick. (From Bewick's *British Birds,* 1789)

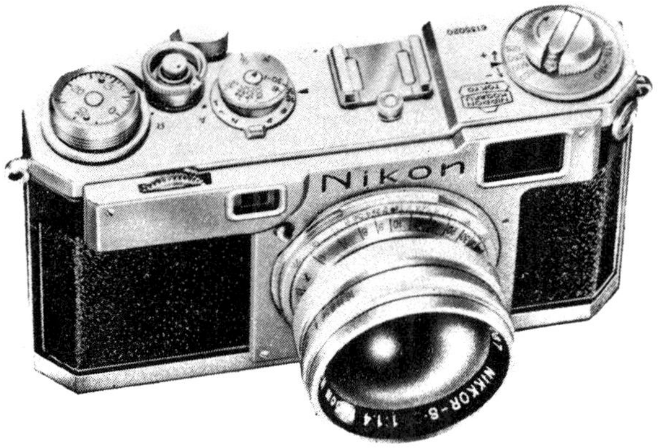

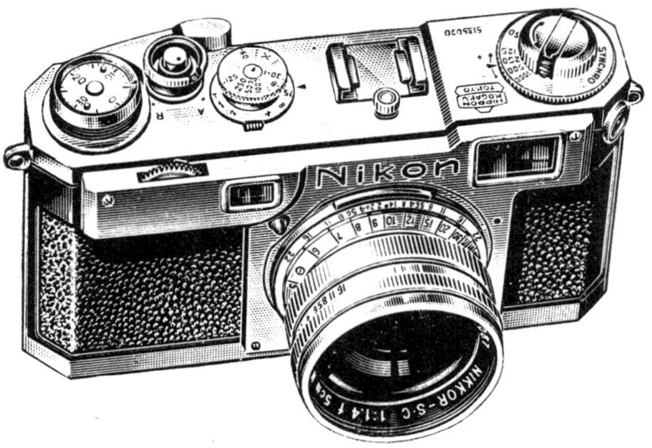

Compared to the photographic halftone on the left, the wood engraving on the right shows greater clarity of line and detail.

spaced lines, and by forgoing the old method of cross-hatching for shading, Bewick was able to create an entirely new kind of woodcut. It rapidly gained wide acceptance and for the next one hundred years or so was used in many countries for illustrating books, magazines, and catalogues. Bewick's own illustrated tomes on animals became famed throughout the world, and the eminent American artist-naturalist John James Audubon, who greatly admired his English colleague, became one of Bewick's close personal friends after a visit to England.

After the invention of photography at the beginning of the nineteenth century, the use of engravings made by photomechanical means soon followed. The first metal engravings made in that way date from 1824 and are credited to Joseph Niepce, whose partner, Louis Daguerre, became famous as the inventor of the daguerrotype, an early form of photograph. Other inventors contributed their own improvements to the process; the English photographer and archaeologist William Henry Fox Talbot had the idea of placing a screen creating a dot pattern between the plate and the film, and thereby invented the "halftone" pattern used to this day in all except a few photomechanical printing processes.

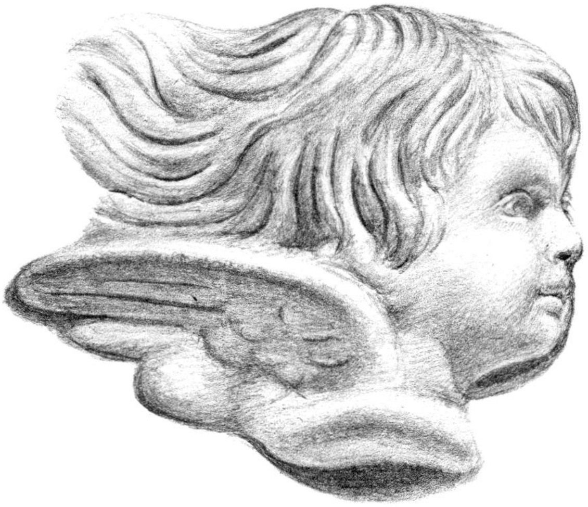

Despite its halftone effect, this pencil drawing on finely stippled film reproduces as line if printed from a contact negative.

By the end of the nineteenth century photoengravings were replacing many of the old hand-tooled wood and steel engravings in books and magazines. Quite a few advertisers, however, continued to prefer the clarity and definition of line of wood engravings to the frequently indistinct and muddy effect of the photographic halftones. Wood engravings were used for a great many advertisements until well into the twentieth century. But then the vastly

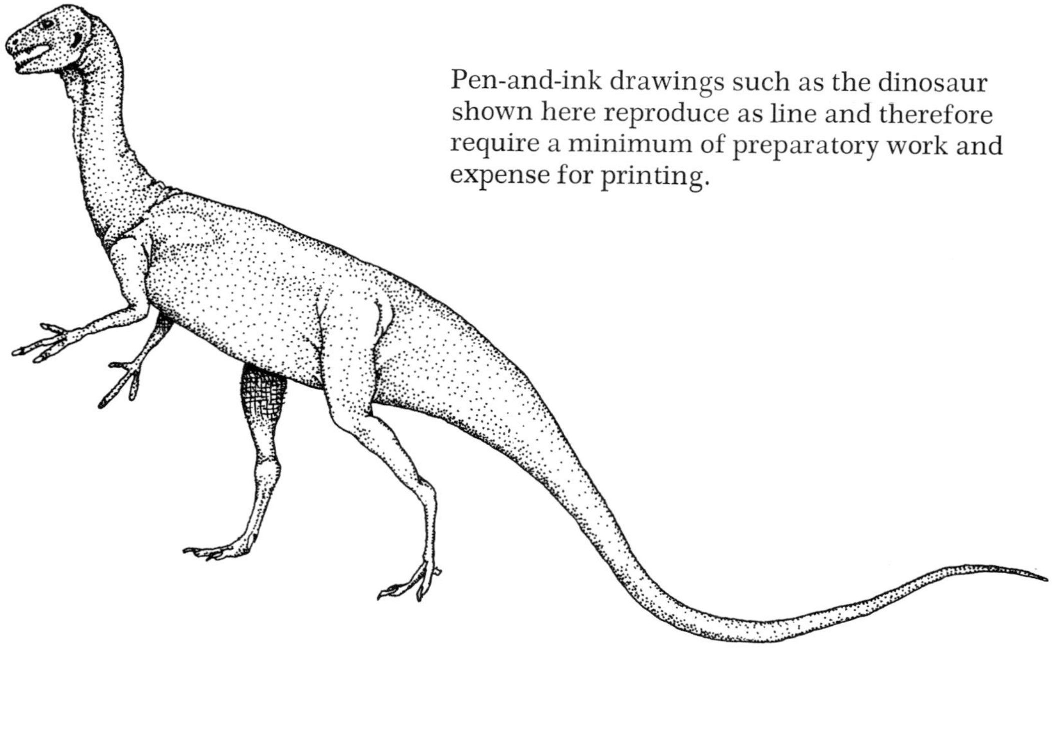

Pen-and-ink drawings such as the dinosaur shown here reproduce as line and therefore require a minimum of preparatory work and expense for printing.

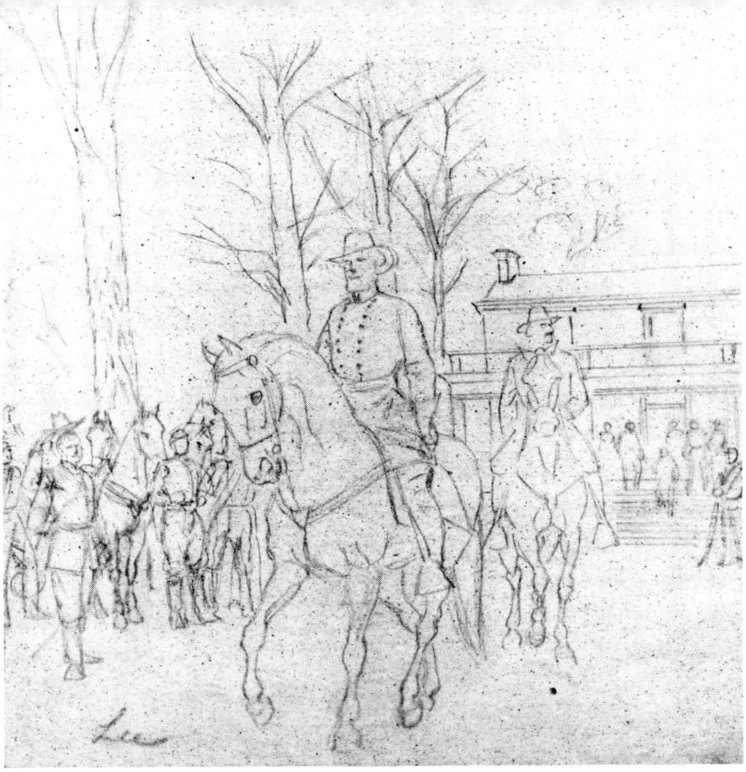

Any wash, or watercolor, illustration such as this one must be turned into a screened halftone if it is to reproduce properly.

improved and highly developed photomechanical process took over almost entirely, for now illustrations could be reproduced not only in black and white but also in color by means of photomechanical separation. Printing from hand-tooled plates of any kind became virtually a thing of the past.

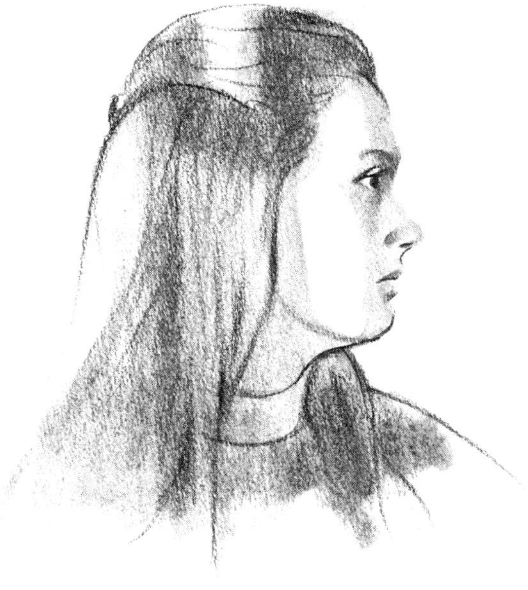

Like most pencil drawings done on paper, this illustration had to be screened and reproduced as halftone in order to preserve the values of the shaded areas.

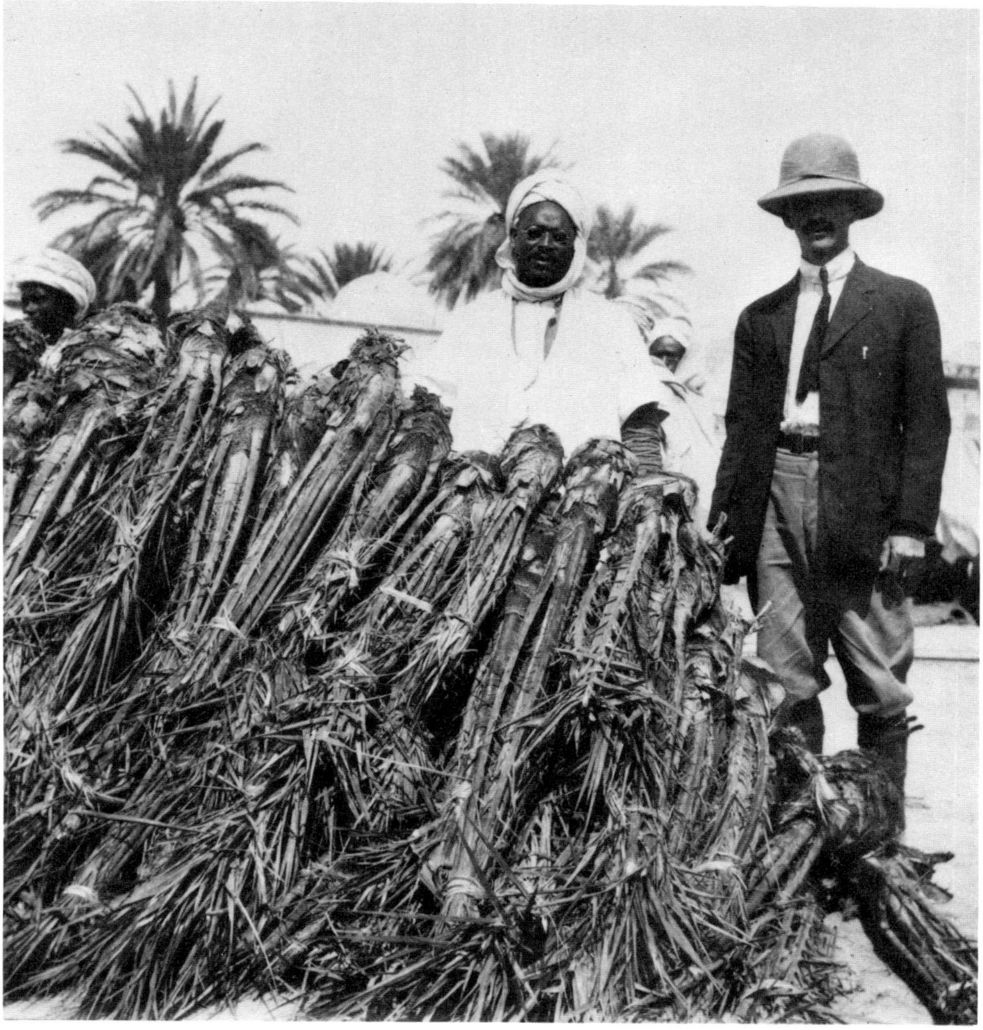

Although halftones respond best to the smooth surfaces of glossy coated papers, this reproduction of a photograph proves that all except the finest screens can be effectively printed on high-quality, dull-coated stock such as the one used in this book.

The beginning of widespread and steadily increasing use of color in printed reproductions can be traced to the discovery of coal-tar dyes in the middle of the nineteenth century, for that discovery permitted the manufacture of colored printing inks on a commercial basis. In the beginning each color had to be prepared separately by hand, for the first photomechanical separation of colored originals was not produced until 1893, and it took some time to perfect the process.

Despite the increasing importance of color reproductions in the twentieth century, black has remained the mainstay of the publishing industry. Books printed in black only far outnumber those featuring color. Black can be used not just for the printing of text matter but also for the reproduction of illustrations ranging from photographs to original artwork executed in a number of different media and techniques. A well-produced book in black and white can be most attractive. The degree of visual appeal depends on a great many factors, including the quality of the original photographs or artwork, the type design, the paper, the quality of the camera work and press work, and last but not least, the kind of black ink chosen for the printing. For although many people tend to think of black ink as being "just black," there are different kinds of black that may be used to suit different subjects.

The least attractive among the black-and-white illustrations are those employing a "dead black" ink on a "dead white" paper. As the word implies, these hues lack life because there is a total absence of chromatic color. The black absorbs and the white reflects every bit of light, thereby producing harsh contrasts usually devoid of visual appeal. However, certain subjects such as technical and mechanical illustrations, where sharpness and clarity of line are important, may actually profit from a starkly black-and-white reproduction.

For the majority of illustrations, though, extreme harshness of contrast is not desirable; the visual appeal of most books is greatly enhanced by some softening of the extremes, and often also by investing the black with a trace of chromatic hues. Depending upon the subject, the choice of an off-white or even a cream-colored paper, plus a black ink containing an admixture of either some red-yellow for a brownish-tinted black or some blue-red for a purplish-tinted black, may do wonders for the attractiveness of the illustrations without increasing the cost. One need only visualize a book on astronomy or space printed in a violet-black ink, or one on ancient artifacts reproduced in a brown-toned black to realize how

An illustration printed in neutral, "dead" black.

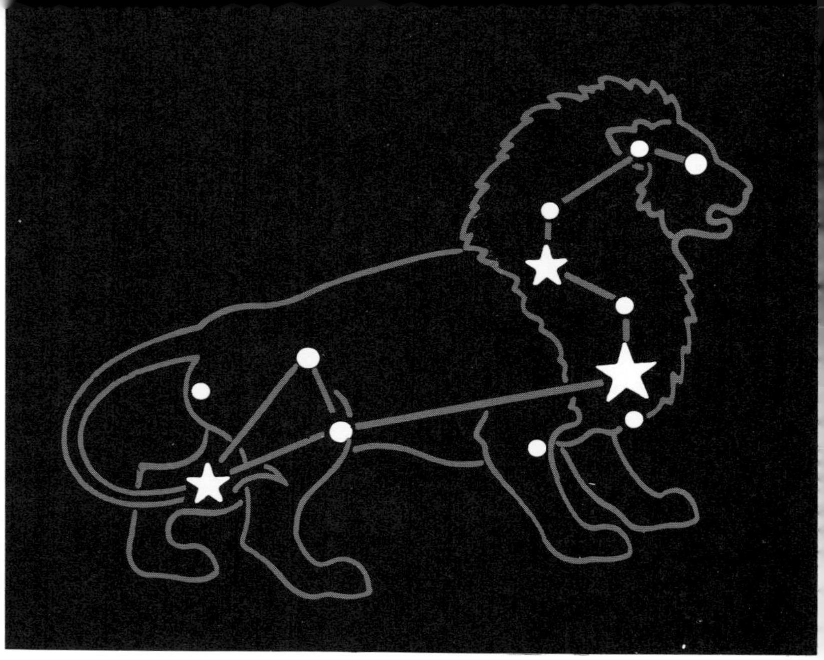

great an improvement such a deviation from the standard black ink may be. Most readers would find these effects pleasing without being consciously aware of the reason, for the receptors in the eye are extremely sensitive and react to visual phenomena quite independently of conscious reasoning.

A well-designed, well-illustrated, and well-printed book in black and white may be a good deal more attractive than one of poor quality featuring full color. Black-and-white books are also, of

An ancient Greek pottery fragment reproduced in neutral black.

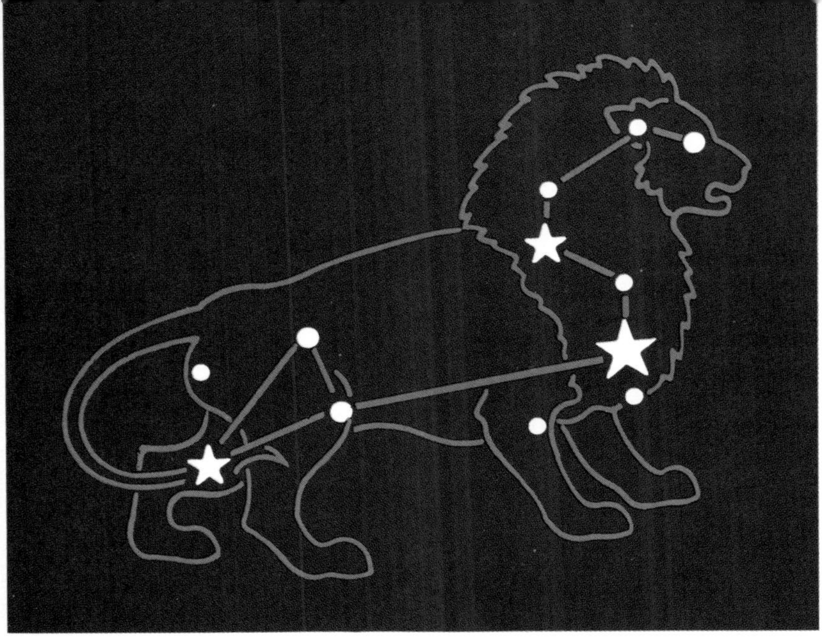

The same illustration printed in a blue-toned black. (Here approximated by adding percentages of process red and blue to neutral black.)

course, much more economical, for full-color reproduction is expensive and is not becoming any less so, despite all the advances and improvements of our electronic, computerized age. For many subjects, colored illustrations are not essential, and so the "basic black," in the past the backbone of all book printing, has once again become the choice for many illustrations that not long ago would have been reproduced in full color.

The same illustration shown in a brown-toned black. (Here approximated by adding tints of process red and yellow to neutral black.)

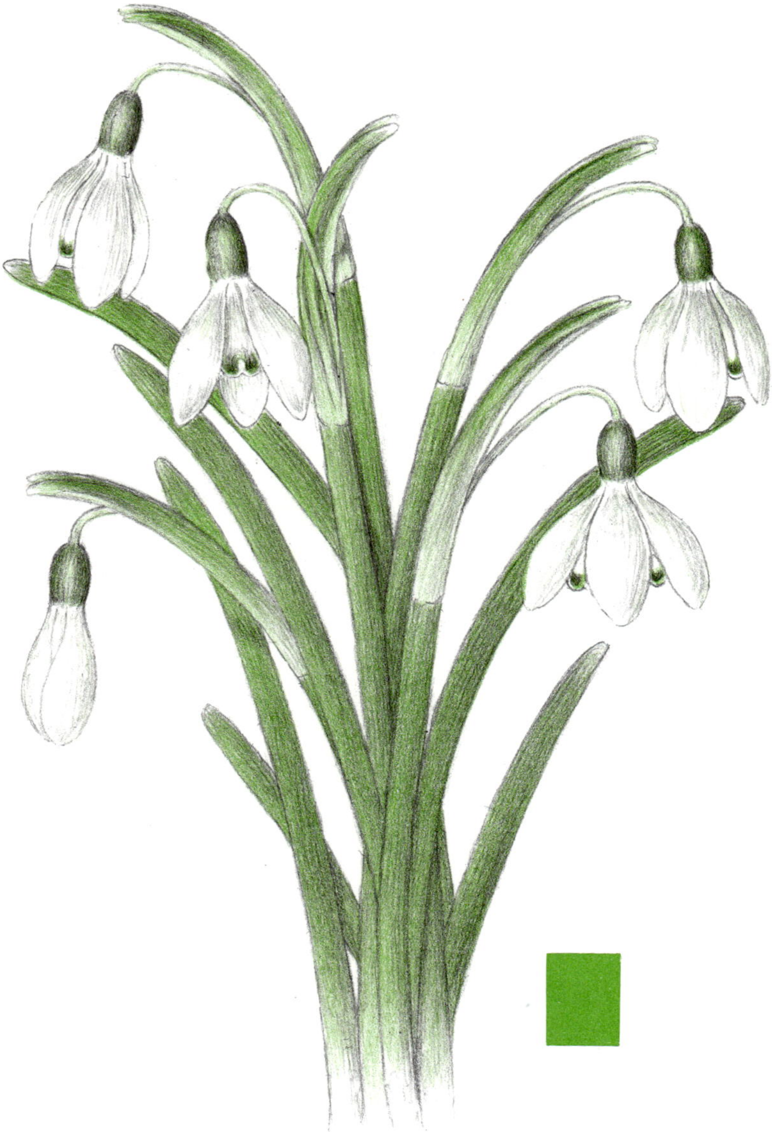

Green makes an excellent second color for illustrations dealing primarily with vegetation such as this outline drawing of snowdrops and their foliage.

Black Plus One Color

Book illustrations are by no means restricted to a choice between black only and full color. The addition of even a single chromatic hue to the basic black may greatly enhance the visual appeal of illustrations for which full color may be deemed not necessary or economically not feasible. So long as photomechanical color separation is not involved, such additions are often well worth the relatively modest increase in printing costs, especially if the books are intended for the young, to whom color in illustrations is important.

Depending on the type of original used and the effect desired, the color may be added in any one of a number of ways. For photographs, the options are limited; one of the most popular is the so-called *duotone,* which is a single-color photograph converted into a two-color halftone by a photomechanical process. The standard duotone is produced by making two images of somewhat differing strengths of the original photograph, and then superimposing the black image over the one printed in whatever second color has been selected.

A much simpler, less costly, and less sophisticated kind of duotone effect may be achieved by simply printing the black halftone image over a flat screen tint of the desired color. Of course, there will be no white areas at all within that illustration, and therefore no highlights. The standard duotone, on the other hand, preserves and often accentuates the highlights of the original photograph, and therefore results in a much more three-dimensional-looking picture.

A third way of adding color to at least some types of photographs is by designating certain areas for solids or screen tints,

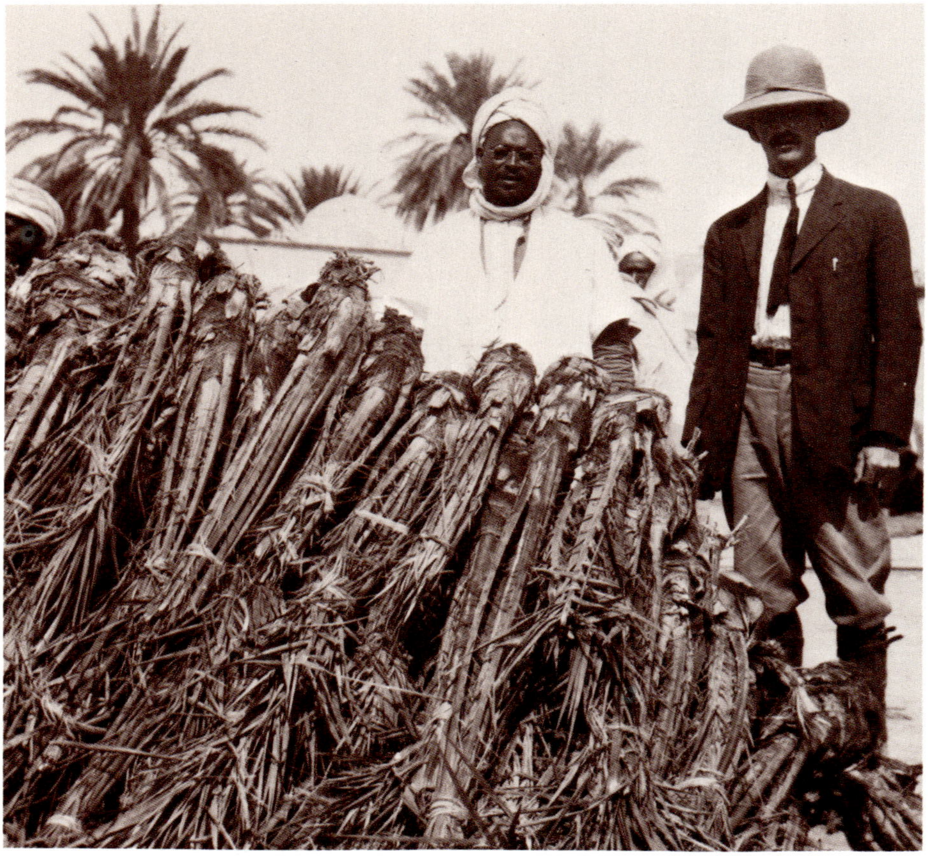

The photograph reproduced in black only on p. 36 has been transformed here into a brown-black duotone by photomechanical means. In this process, the original's highlights are preserved.

which later are filled in by the printer for an impression in the second color. Either such areas are opaqued out on the black-image negative so that they will print only in color, or the black is overprinted on the color for a mixture of both; a combination of the two methods may also be employed. Thus a scene showing an expanse of water and sky could be handled by replacing the light black halftone of the sky with a deep tint or even a solid blue and overprinting the black of the water area on a light blue tint for a variety of hues.

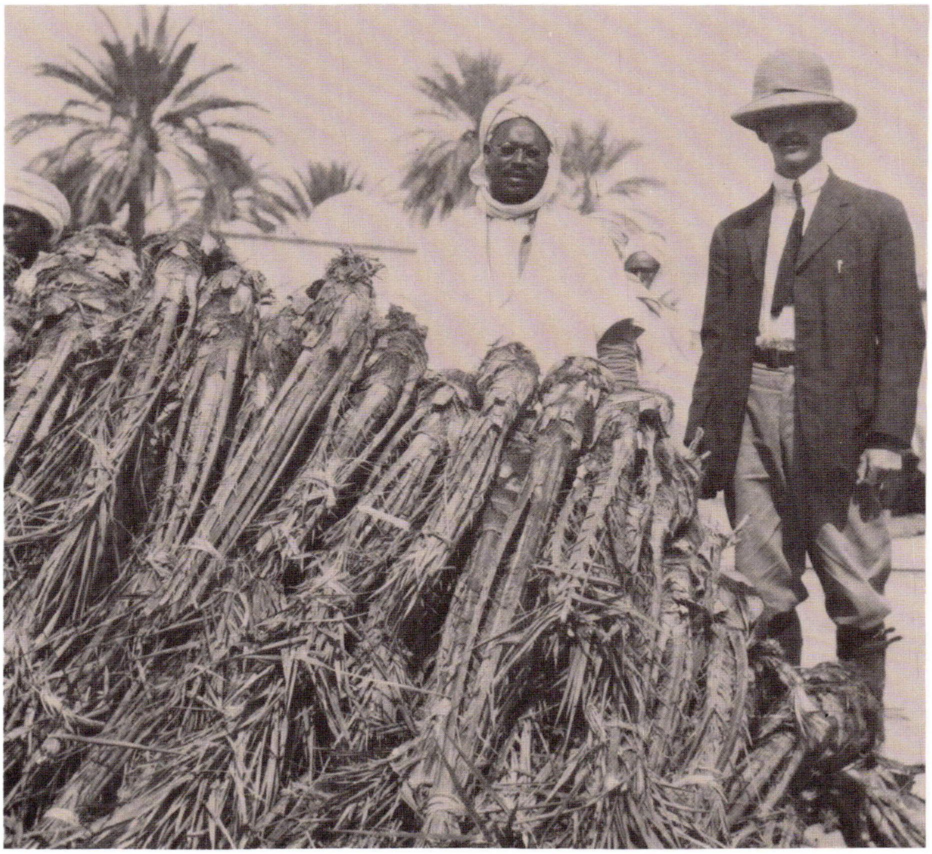

Overprinting the black halftone image on a 30% screen tint block of the same brown creates a simpler and less expensive duotone effect. However, comparison with the duotone on the left shows that the flat color screen diminishes the highlights and depth perspective of the picture.

If the book deals with a specific subject, such deliberate "color accenting" of pertinent objects may create a definite color theme throughout.

Color may also be added to photographs in hand-rendered separations, but this technique is less often used than any of the others because it has to be done skillfully in order to look good. In some instances, however, such hand-rendered color additions may be important for imparting more definition to specific objects not clearly discernible in the black image, for picking out highlights, or for providing a suitable background.

Technically at least, the mechanical ways of adding color to photographs just described apply equally to hand-rendered black illustrations. In practice, however, these methods are hardly desirable in hand-rendered illustrations because the color would be merely an afterthought, an embellishment. In any illustration *planned* for two colors from the start the second color is an integral part of the composition. Such a drawing not only handles the black differently but also allows for greater freedom and a wider range of possibilities in the use of the color, and so usually produces a more satisfying effect.

The second color may be made to serve a variety of purposes—as a background, for highlighting, or even as the basic color with

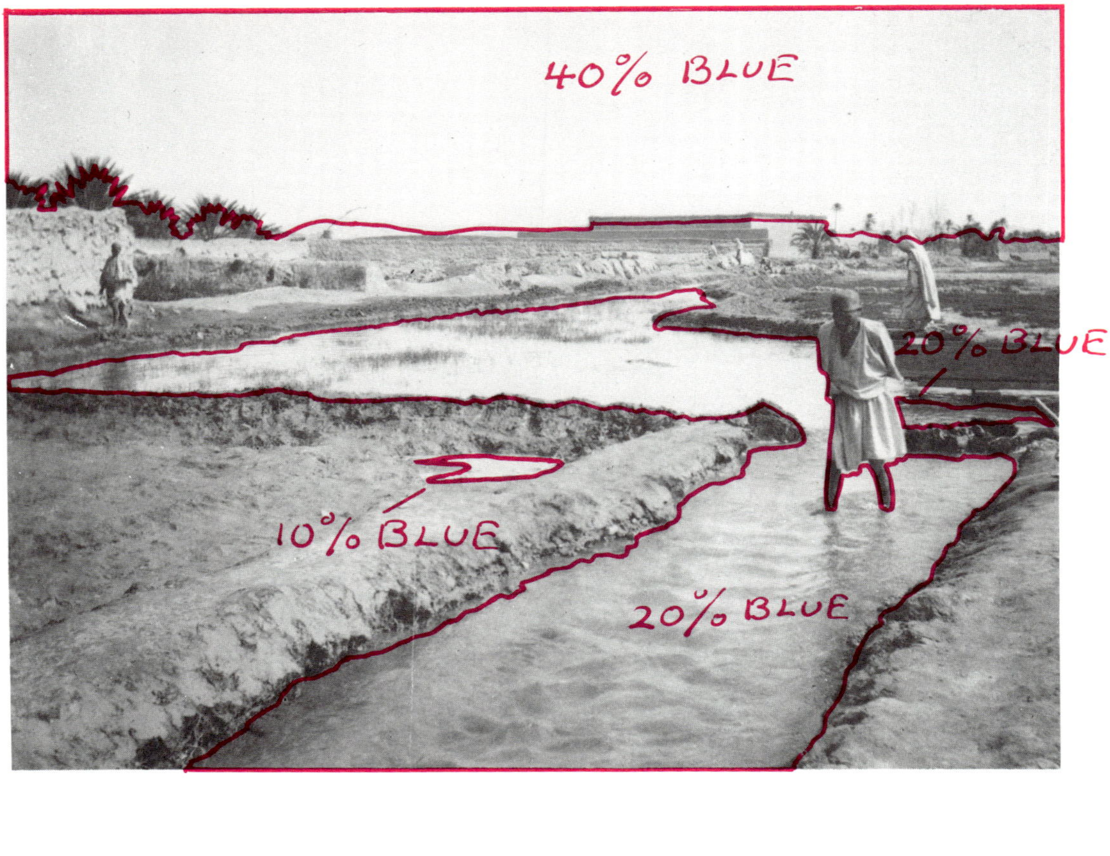

black used only as a strengthener or for a "skeleton" outline. The color may be added in the form of solids, halftones, or a mixture of both, depending on the author's and the illustrator's conception of what it is supposed to accomplish.

By and large, a mixed hue is a better choice than a single-tone primary for any one-color supplement to a black illustration. The multiple color permits a richer variety of hues both by itself and in admixtures with black. A warm golden brown, for example, or a slightly olive-toned green may be used for a whole range of different shades from pale, yellowish-looking tints to rich dark browns or

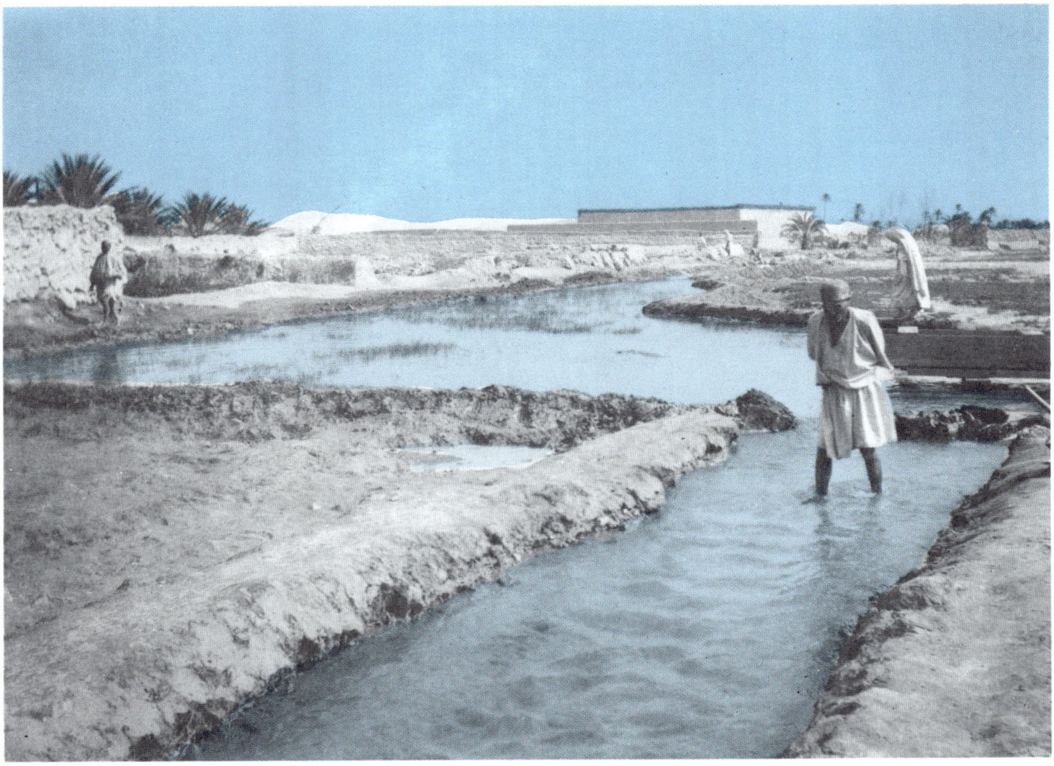

Solids or screen tints of a second color may be added to any black illustration by marking the pertinent areas for color percentages on an overlay (left). The printer fills in these areas with the appropriate tint strengths on the color plate and then reproduces the illustration in two colors

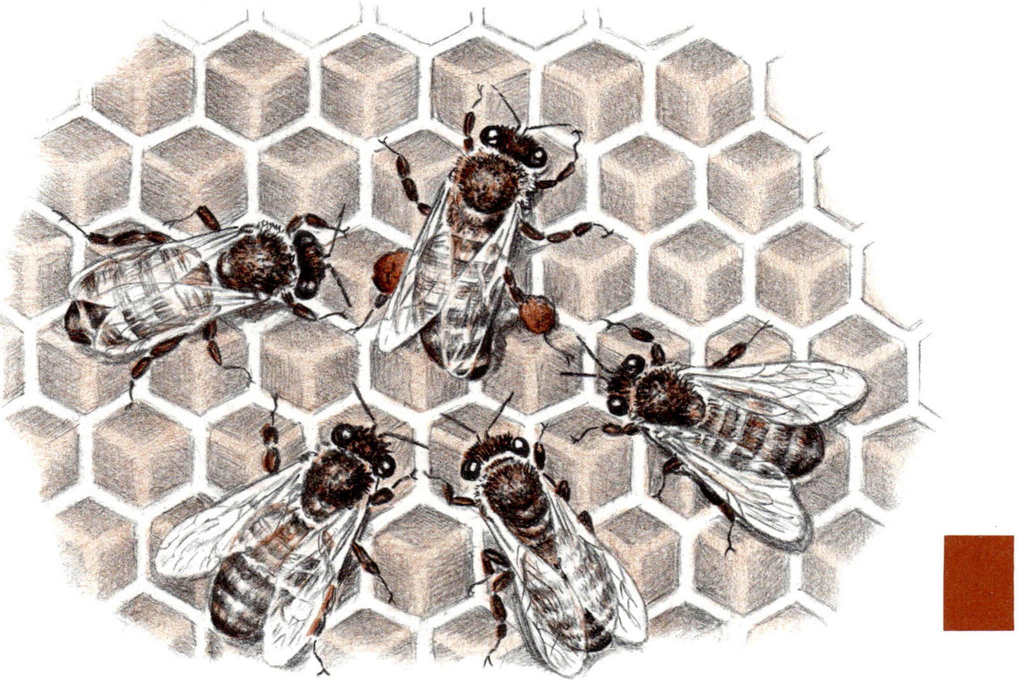

Used as tints for the background and for the insects, and full strength for the pollen baskets, warm brown was a perfect choice for the second color in a book on bees and wasps.

olives when the color is used full strength and overlaid with tints of black. The choice of the second color naturally depends largely on the subject. Illustrations of plants or animals lend themselves to additions of green or brown, respectively, whereas a book of fairy tales or other fictional accounts for young children might well be enhanced by such hues as purple, bright red or orange, or peacock blue.

Art techniques used in preparing the color separations may vary as widely as those employed for the black illustration, nor does the separation have to feature the same medium used in rendering the black drawing. A black line illustration executed in pen and ink, for example, may be supplemented by a color separation done in wash or pencil and reproduced as a halftone. At the other end of the scale, halftone drawings frequently benefit from color added in the form of solids, and some black wash illustrations are very effectively colored by flat screen tints marked for percentages on the separation by the artist and filled in by the printer. Whatever the technique used, experienced and skilled illustrators will be able to extract maximum effectiveness and appeal from that one additional color at their disposal.

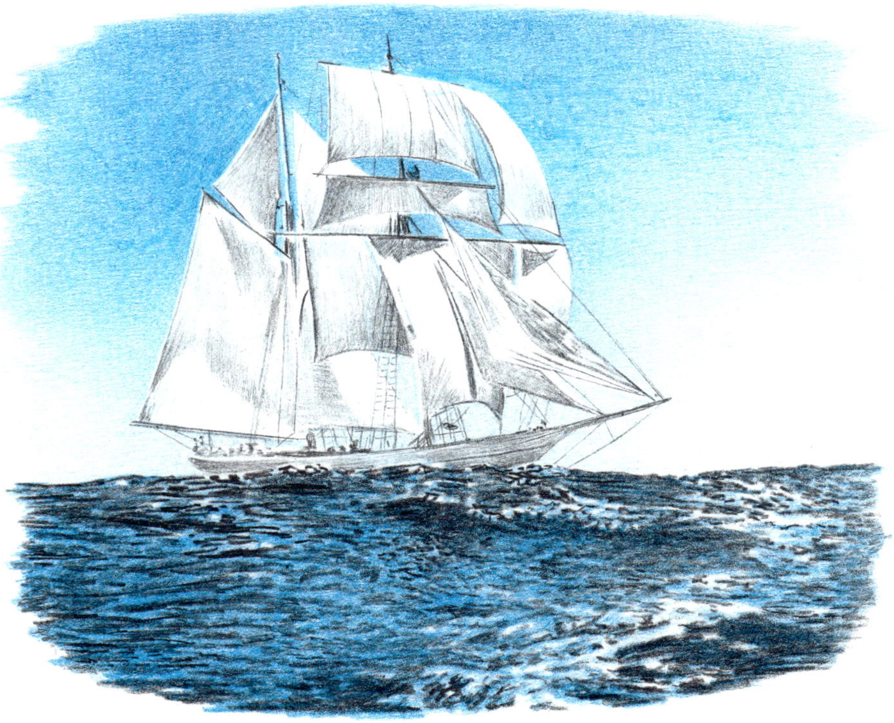

Any second color may be used in a number of different ways. In this illustration of a sailing ship, the blue is in fact the main color, with black serving to outline, strengthen, and delineate specific detail.

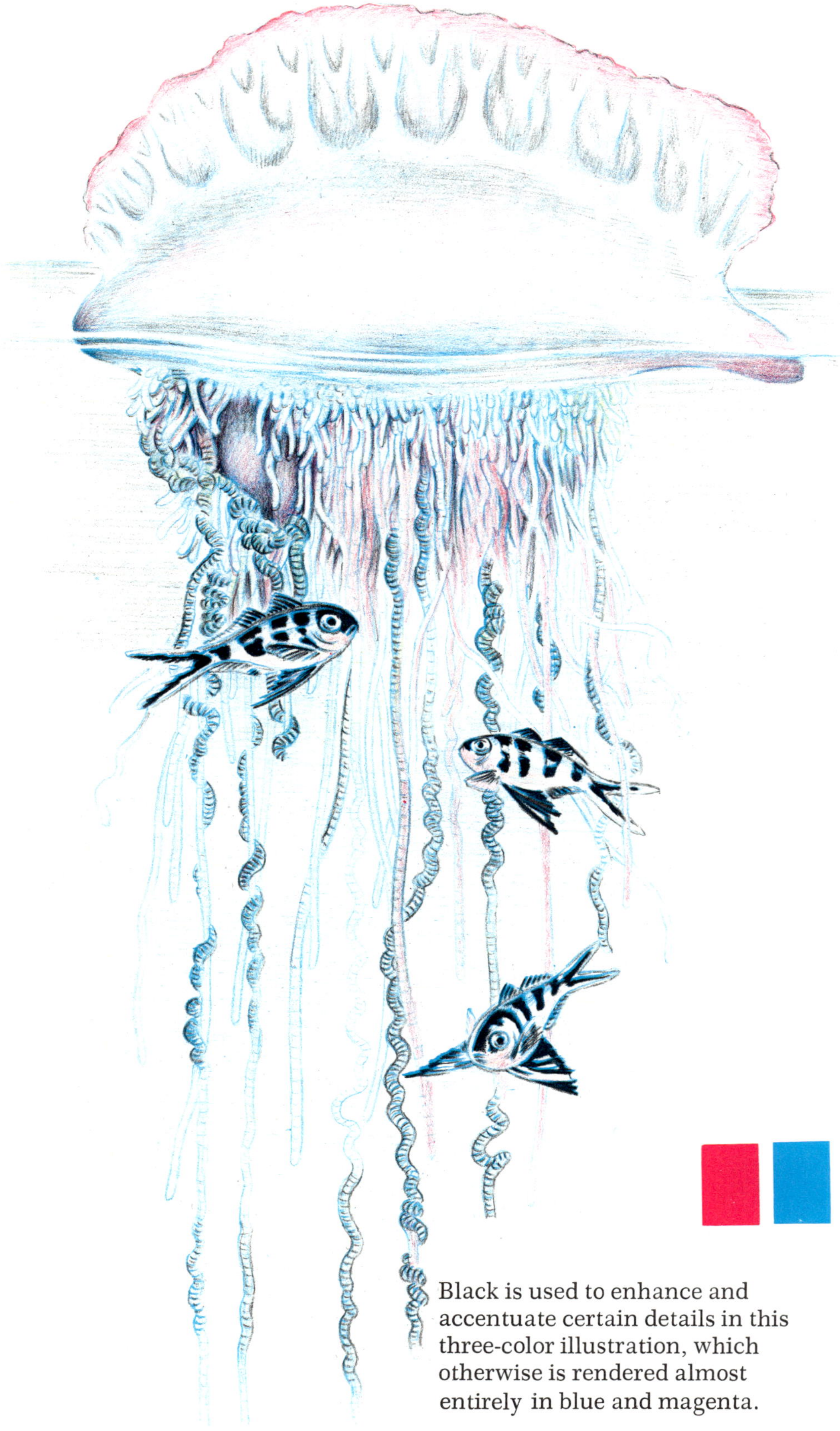

Black is used to enhance and accentuate certain details in this three-color illustration, which otherwise is rendered almost entirely in blue and magenta.

Black Plus Two Colors

Not many books call for illustrations featuring two colors in addition to black, and the majority of those that do are in the juvenile field. Bright, attractive pictures rendered in three colors can be employed very successfully to complement stories and fairy tales, especially those aimed at the very young. There are, however, a good many other possibilities for the use of three-color artwork.

Whatever the subject, such illustrations usually have to be hand rendered *and* preseparated. For that reason, photographs are rarely considered; there is no practical mechanical process by which a single-color photograph can be turned into a three-color image. If such an effect is desired, the colors have to be added by hand on individual separations. This means a combination of at least two different techniques and demands great skill if the effect of the photographic image is to remain unimpaired. Nevertheless, there are instances when subtle coloring can greatly enhance black photographs. Colors may also accentuate important detail or make it easier to distinguish between various objects in the picture that are indistinct in the black original.

Because the black photographic image has to be reproduced by a screened halftone, matters are simplified if the colors are applied as line renderings, either as solids if the proposed hues are pale, or in simulated halftones that produce delicate shadings and tints though they are rendered in line. Separations designed for halftone production, such as watercolor or flat screen tints, are, of course, perfectly possible. However, the printed result may not always warrant the additional expense of three-color halftone reproductions.

A satisfactory result is much more likely if the illustrations are done entirely by hand, and have been planned for three pre-separated colors from the beginning. As in the case of two-color illustrations, the separations do not necessarily have to feature the

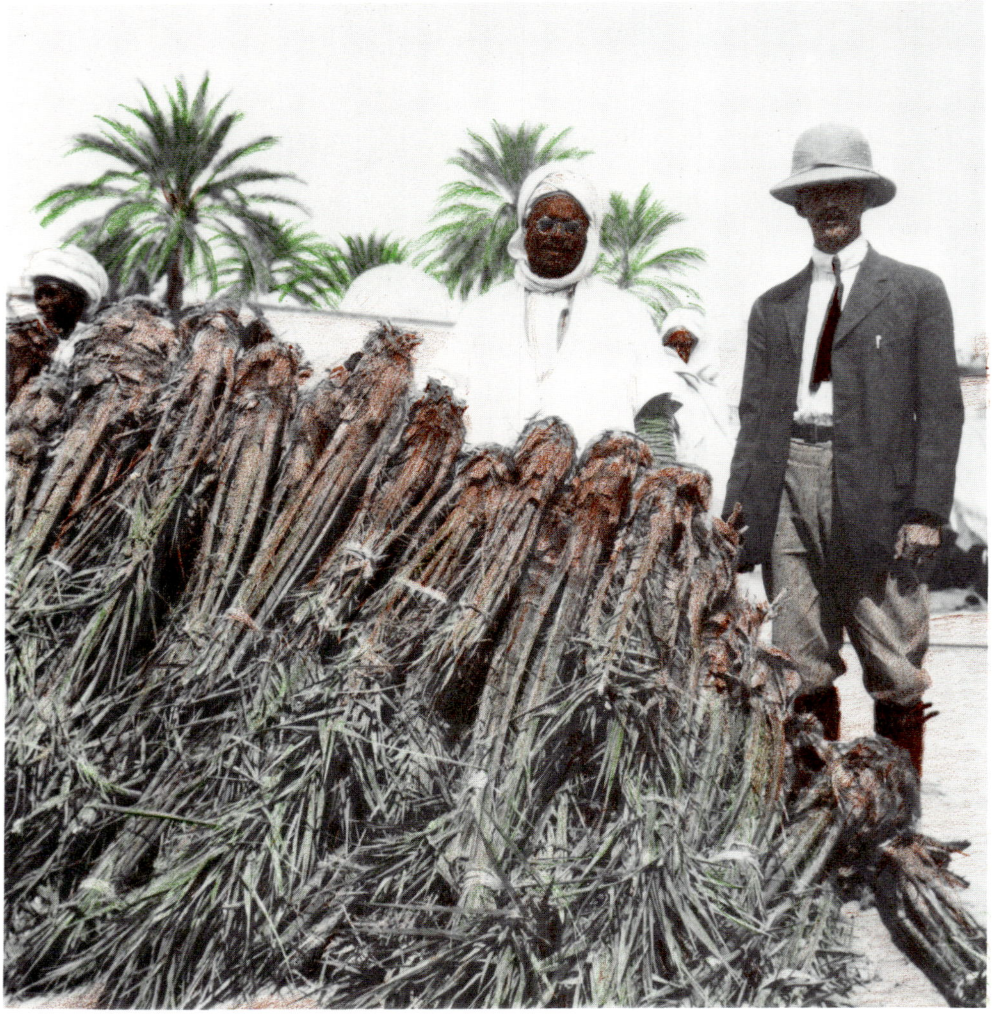

The halftone reproduced in black on p. 36 and as a duotone on p. 42 here has been converted into a three-color illustration by supplying it with line overlays printing in green and brown. Both hues are combinations of differing percentages of all three pigment primaries.

same medium and technique as the black drawing. However, any mixing of different media such as pencil, ink, and paint on the same separation will work very strongly against a satisfactory reproduction. The camera lens records only what it "sees"; it does not have the human eye's faculties for delicate adjustment and compensation. That is one reason why an illustration made for reproduction differs so fundamentally from a work of art meant for framing. The latter is the end result, the former the starting point of a long mechanical process that culminates in the printed image. It matters little how good the original looks if it reproduces poorly, and even the best printers are not magicians. Artists who mix different media and pile on the paint when making an illustration—especially a preseparated one—create a multitextured and multilayered image that may be very difficult to re-create adequately on film by photography. Such artists have only themselves to blame if the final reproduction turns out to be a disappointment.

Delicate tints of multi-toned green and brown add life and detail definition to this originally all-black, pre-World War I photograph.

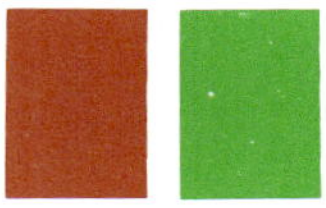

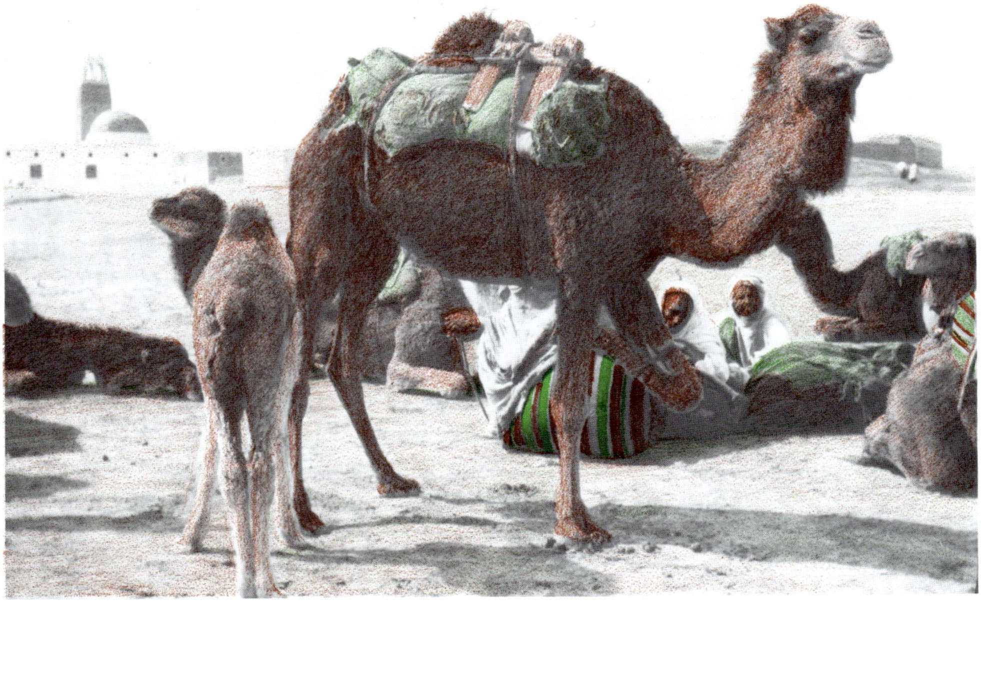

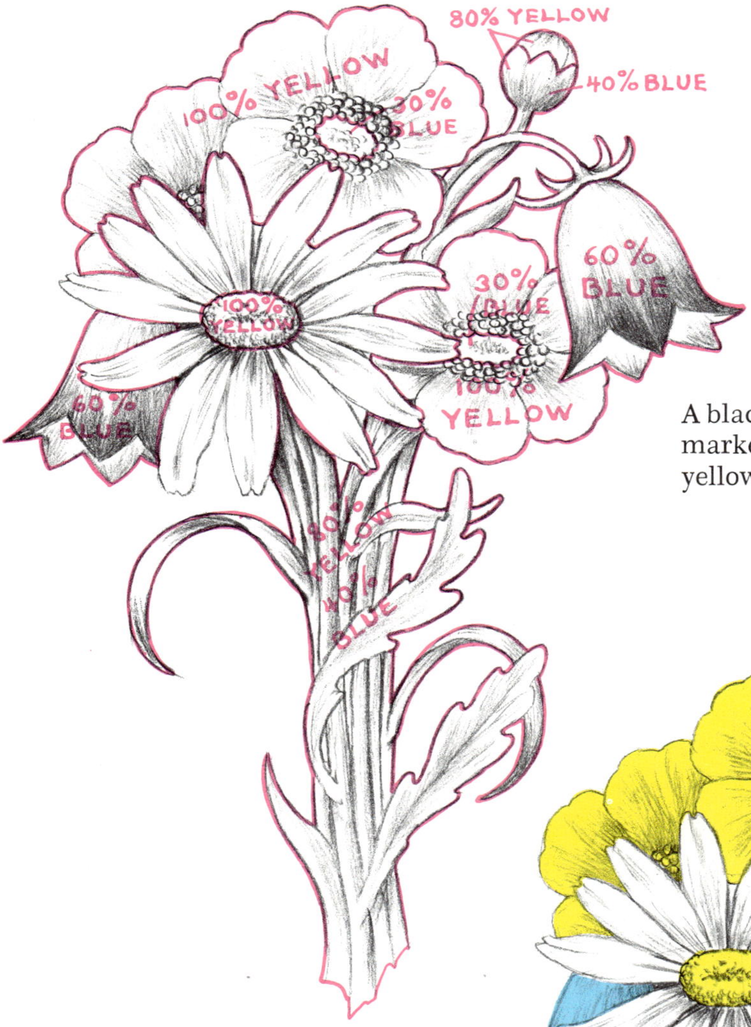

A black line illustration has been marked here for flat screen tints in yellow and blue by the illustrator.

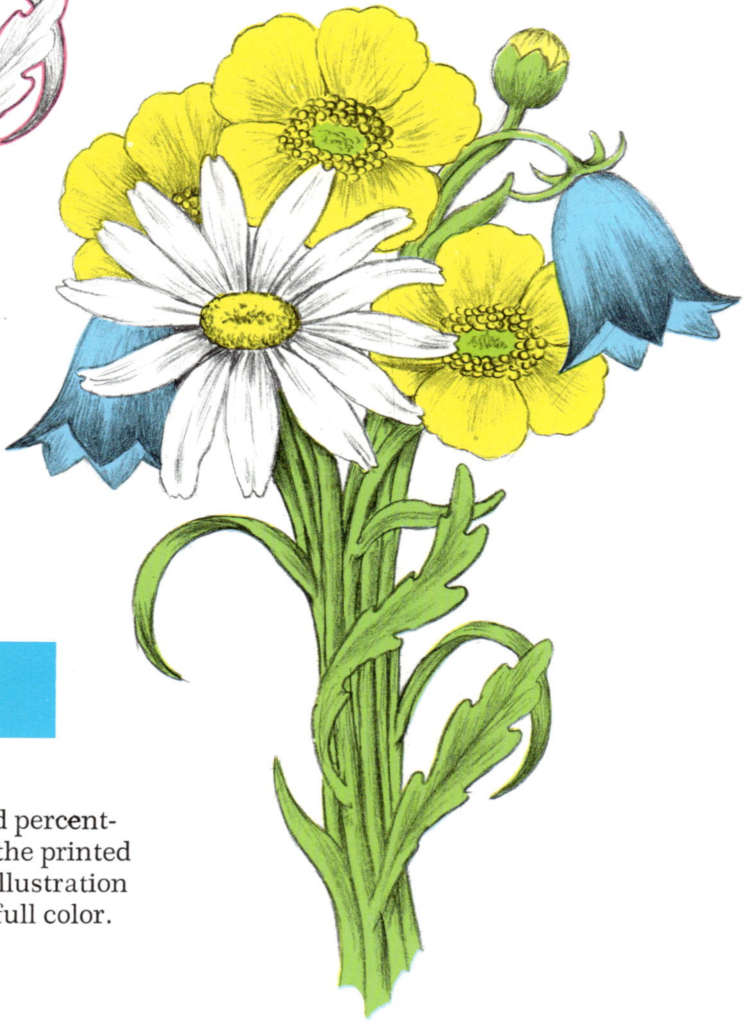

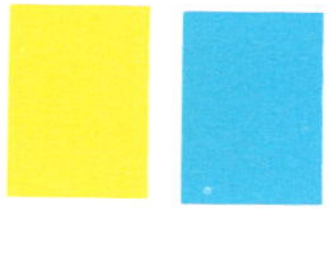

Filled in with the specified percentages by the lithographer, the printed image of this three-color illustration produces a semblance of full color.

Apart from observing a few basic rules, however, three-color preseparated artwork offers considerable freedom not only in choice of techniques but also in ways of using the two additional colors to maximum effect. If selected with care and forethought, two colors can produce a great range of hues and combinations for an attractive result. There are no hard and fast rules for the choice of colors because different subjects call for different treatment. One should keep in mind, though, that the range of combinations will be somewhat limited if two pigment primaries are selected. Thus magenta and yellow can yield only orange-red hues, whereas a choice such as orange-yellow and greenish-blue is more versatile. In addition to the full-strength hues, they can substitute for yellow and blue in the lighter shades and also combine for a wide range of

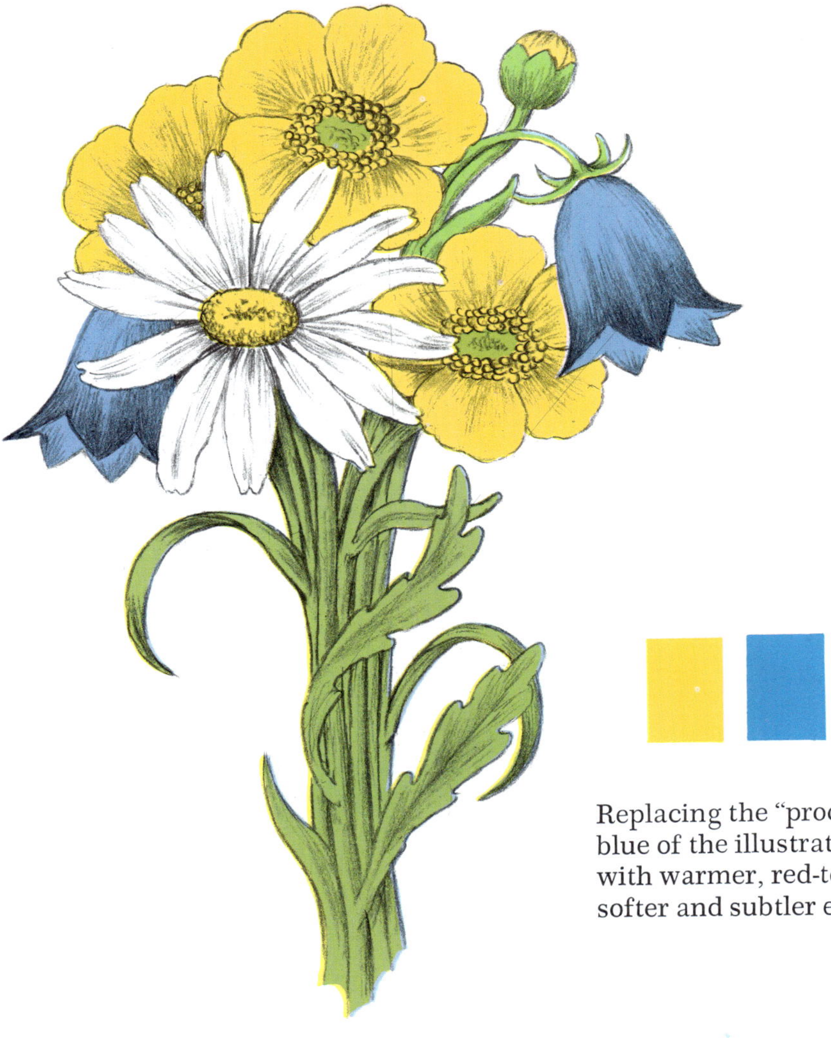

Replacing the "process" yellow and blue of the illustration on the left with warmer, red-toned hues yields softer and subtler effects.

browns and olive greens. With a judicious admixture of black, the spectrum of different hues can be increased to a point where the illustrations might almost be mistaken for certain types of four-color artwork.

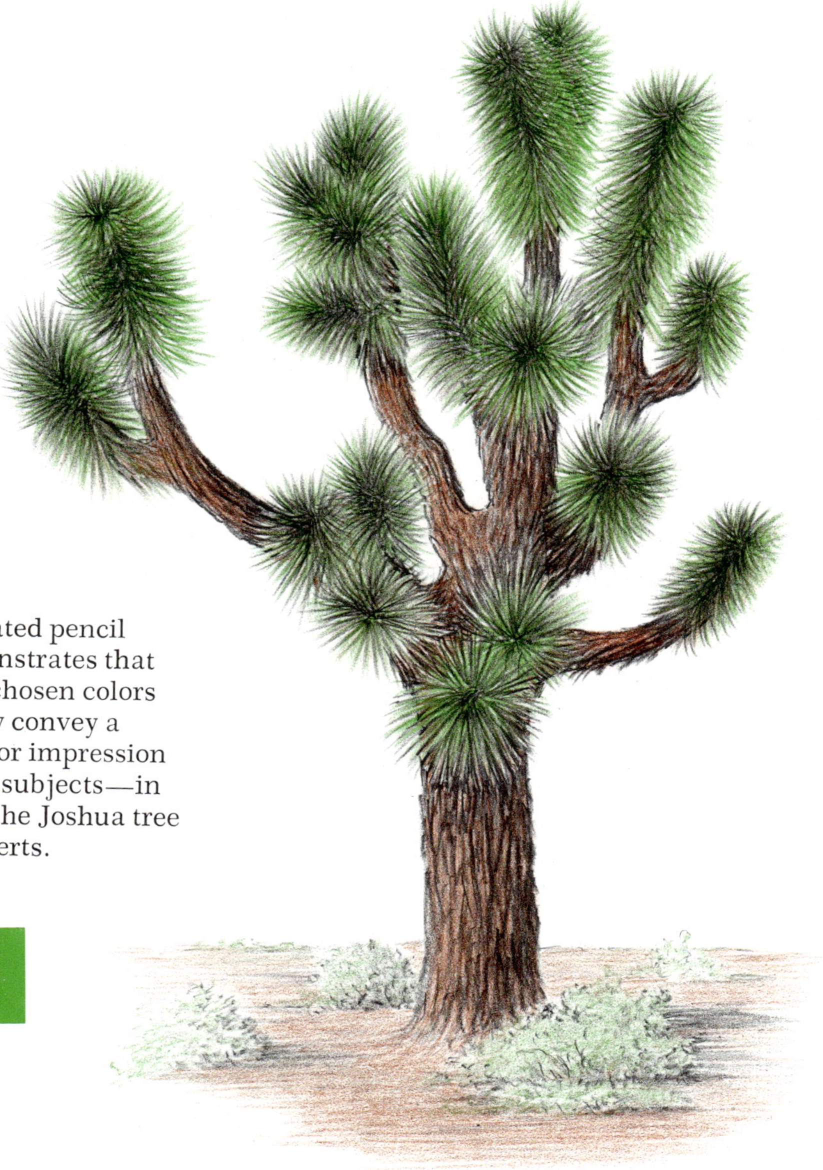

This preseparated pencil drawing demonstrates that two carefully chosen colors plus black may convey a virtual full-color impression of appropriate subjects—in this instance, the Joshua tree of western deserts.

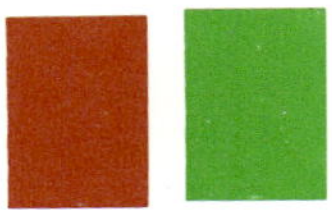

Indeed, two colors plus black may well take the place of full color in some instances because there are subjects that do not require the full range offered by the three primary colors. A book on desert life, for example, could be illustrated quite adequately by using a warm brown and an olive-toned green in addition to black. Some aquatic environments such as a swamp or a pond might be successfully reproduced in three colors by choosing an ochre yellow and a blue, from which both greens and browns may be fashioned for a fairly wide range of hues.

Two colors plus black does not quite add up to "full color," but an artist can find many ways of using them imaginatively and for a variety of effects to create colorful and attractive illustrations.

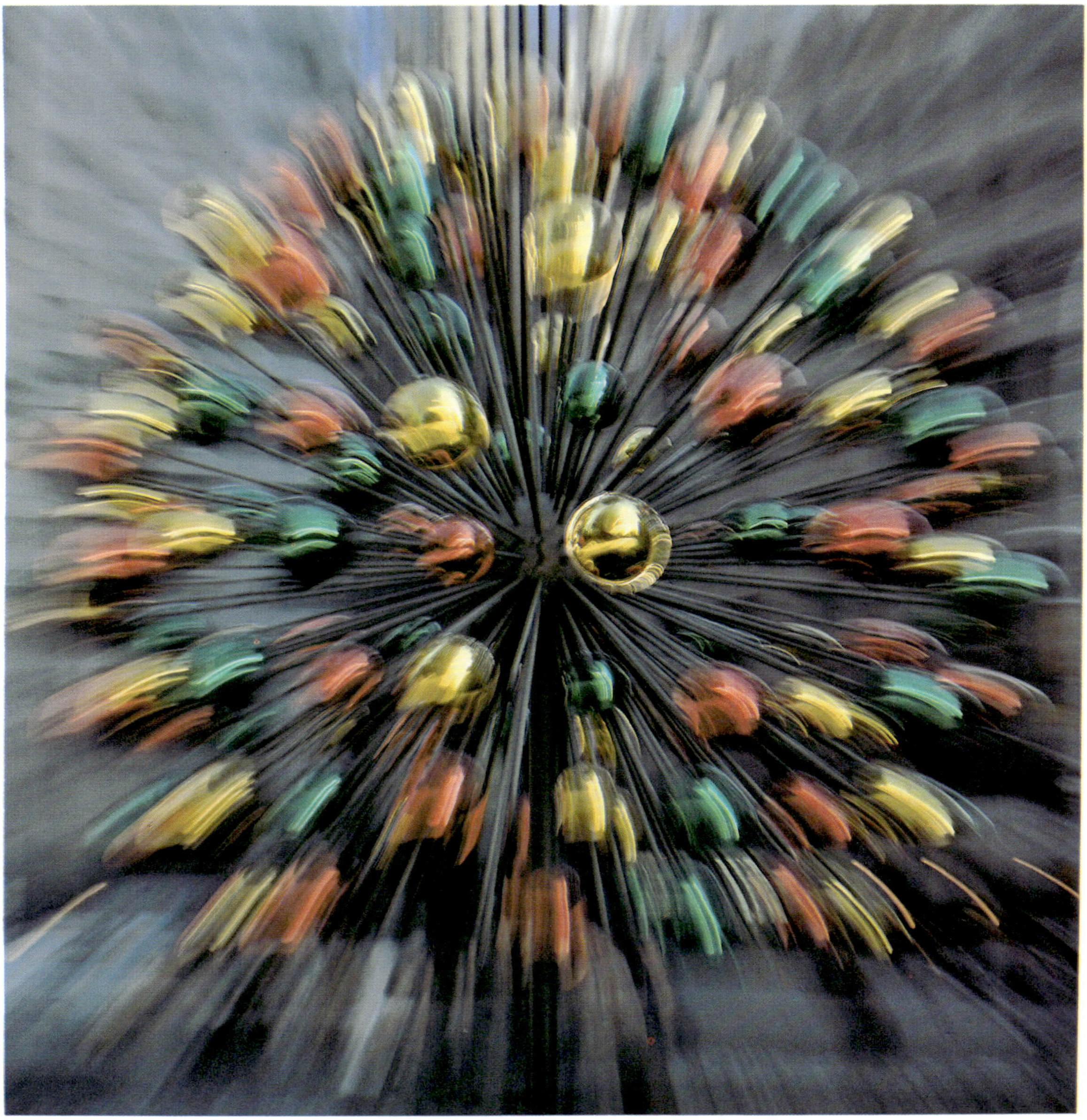

Standard four-color process reproduction of a color photograph whose starburst effect was achieved by deliberately zooming the lens during exposure.

Four Colors for Full Color

The establishment of the printing process in Europe in the fifteenth century heralded a steadily accelerating change in the general availability of books. This meant not just reading matter, but also —and often especially—the illustrations that accompanied the text. Although early books dealt largely with religious subjects, other types of publications soon became popular. In time, illustrated texts ranging from nature to art and architecture found an ever-widening circle of readers. Until about two hundred years ago, however, all illustrations, regardless of what they depicted, were printed in black, and about the only way to have color in such reproductions was to add it by hand.

Methods of making colored imprints were slow to develop mainly because of the lack of suitable colored inks. When such inks finally became widely available, the frequency of color printing took a sharp upturn. Until fairly recently all color plates had to be manually prepared. Skilled craftsmen tooled and etched the areas designed for color into the appropriate plates, or drew them on special stones. Lithography, or "stone writing," became one of the most popular and widely used printing techniques in the nineteenth century. A process invented by the German Aloys Senefelder, lithography used porous stones on which designs were traced with a greasy pencil. These stones were then washed with a special solution that made the fatty substances insoluble, and were finally thoroughly wetted down. Since inks will not adhere to wet surfaces but have an affinity for greasy substances, the design traced by the grease pencil on the stone was the only area that retained the ink and so produced a printed image of the tracing when a sheet of

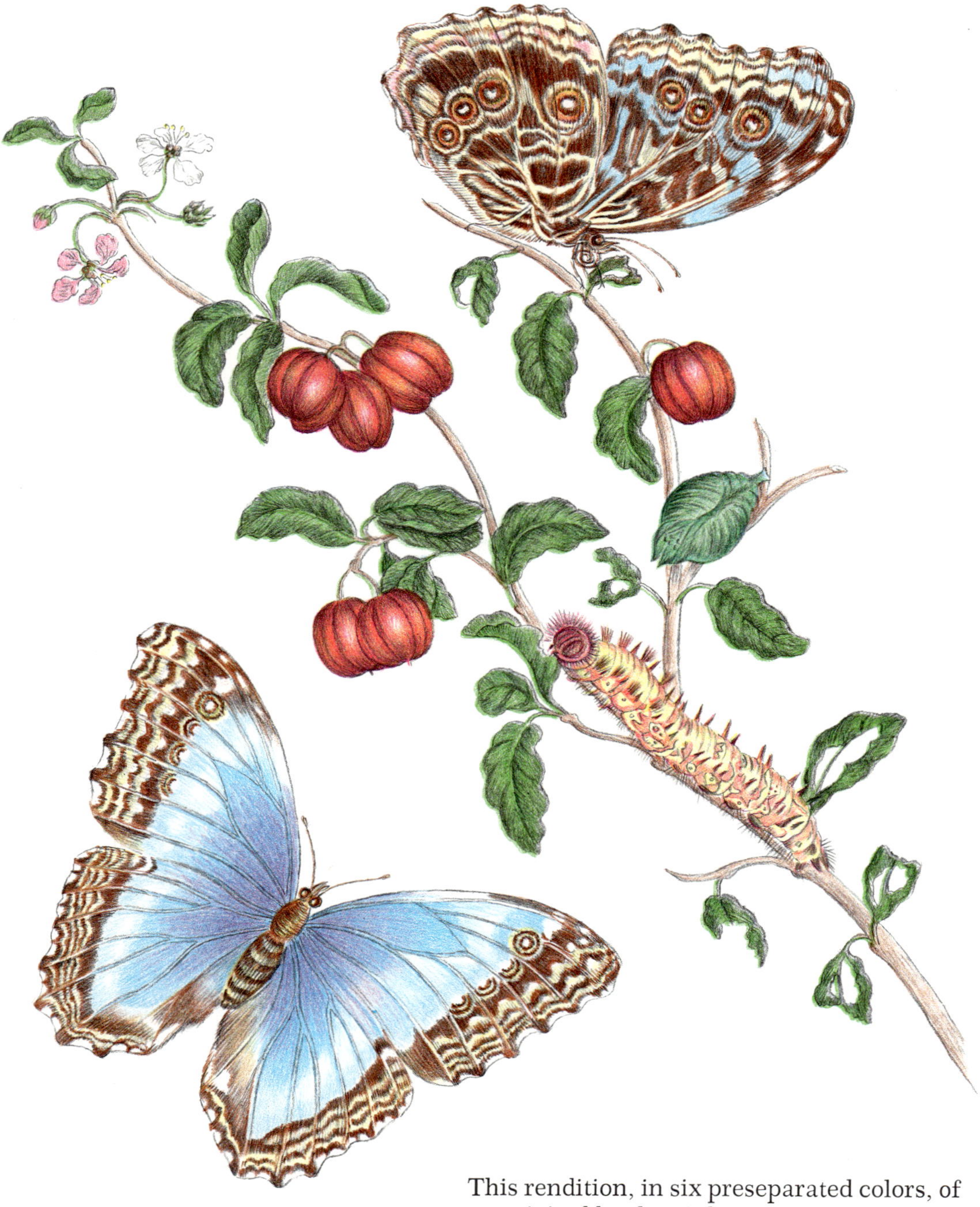

This rendition, in six preseparated colors, of an original by the 17th-century German artist Maria Sibylla Meriam, resembles some of the multi-colored lithographs of the 19th century.

paper was pressed against it. Lithography thus differs from all kinds of engravings by relying not on raised versus depressed areas, but rather on chemical reactions to produce a plate whose image is all on the same level.

Until the end of the nineteenth century most reproductions featuring color were made by the original lithographic method. Highly skilled craftsmen who had to be trained for years prepared the stones for the various color "plates." Many of these old reproductions were not limited to our present standard four colors; printing in six, eight, or even ten different hues was not at all unusual, and that fact accounts for the beautiful delicate shadings of many old lithographs. Today the use of more than four colors in book printing is exceptional and limited to very special editions.

The successful introduction in 1893 of the photomechanical process for reproducing full-color originals revolutionized the printing industry. At the same time, a more or less accidental invention signaled the beginning of the end for the old-style lithography employing stones, and instead started what was to become known as offset lithography. The new process depended on the same type of chemical reaction but replaced the stone of the old lithography with a thin metal plate used to *offset* an image on a rubber mat, which in turn printed the image on a sheet of paper. It was found that the soft rubber tended to produce a cleaner, sharper image, especially in long press runs, than one made from a hard metal or stone surface. In the twentieth century offset lithography gradually became the most widely used method for book printing.

In order to understand the process of four-color separation by camera, it is necessary to recall the optical principles of primary/secondary color relationship, for all photomechanical separation is based on the trichromatic theory of light. The process of isolating the component colors of a full-color original picture in such a way that it may later be re-created in a printed image with the help of certain ink pigments begins by photographing the original three times, each time through a filter of a different color. The filter hues correspond to those of the light primaries red, green, and violet-blue, although the last-named filter is usually referred to as just "blue" or "dark blue." Some filters are designed to transmit light of only a single wavelength; one blue filter used by the world-famous Zeiss Optical Works in Germany was geared to precisely 437 mμ. It was made of two specially prepared sheets of glass colored with methyl violet and Berlin blue. By transmitting only monochromatic light that represents, as closely as possible, the established wavelengths of the individual light primaries, the filters used in process color separation affect the end result, for the degree of the filters' color fidelity largely determines the quality of the separations.

By photographing a full-color picture through a filter that transmits only a single light primary color, exactly one-third of the picture's reflected rays are registered on the resulting negative. A blue filter, for example, permits the negative registering, on film, of all the violet-blue in the original. This is the first of several steps whose net effect is the removal of the blue and the registering of the original's red and green content on the photographic sep-

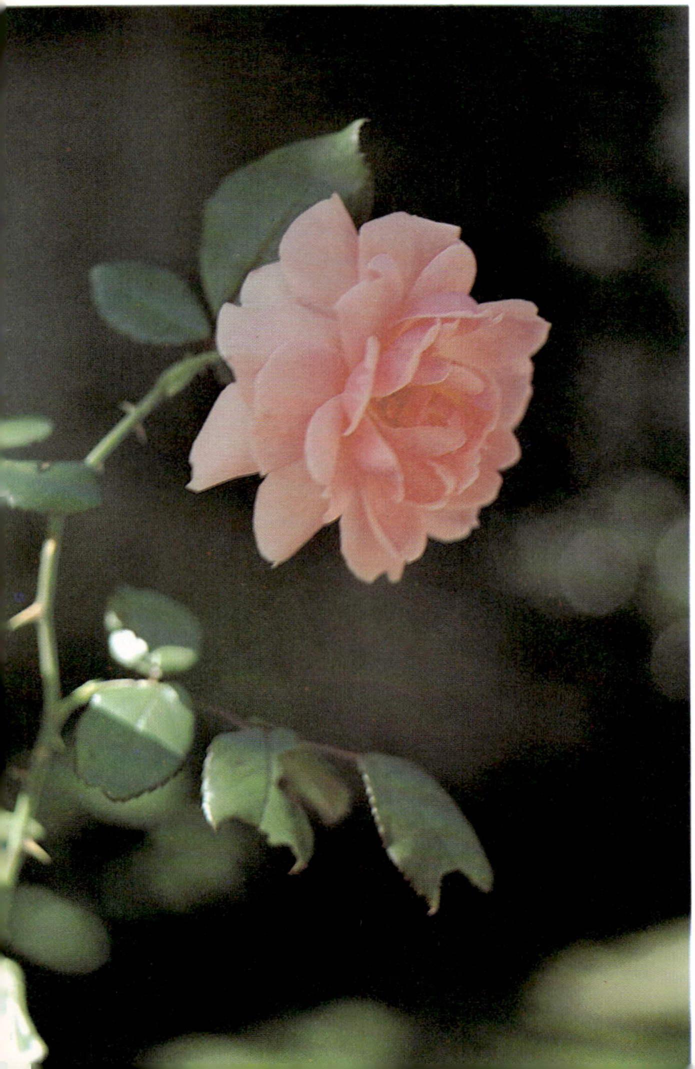

Process reproduction of a rose photographed against a dark, out-of-focus background. (Compare this imprint on the dull-coated paper of the book with that of the same halftone reproduced on the coated and laminated stock of the jacket.)

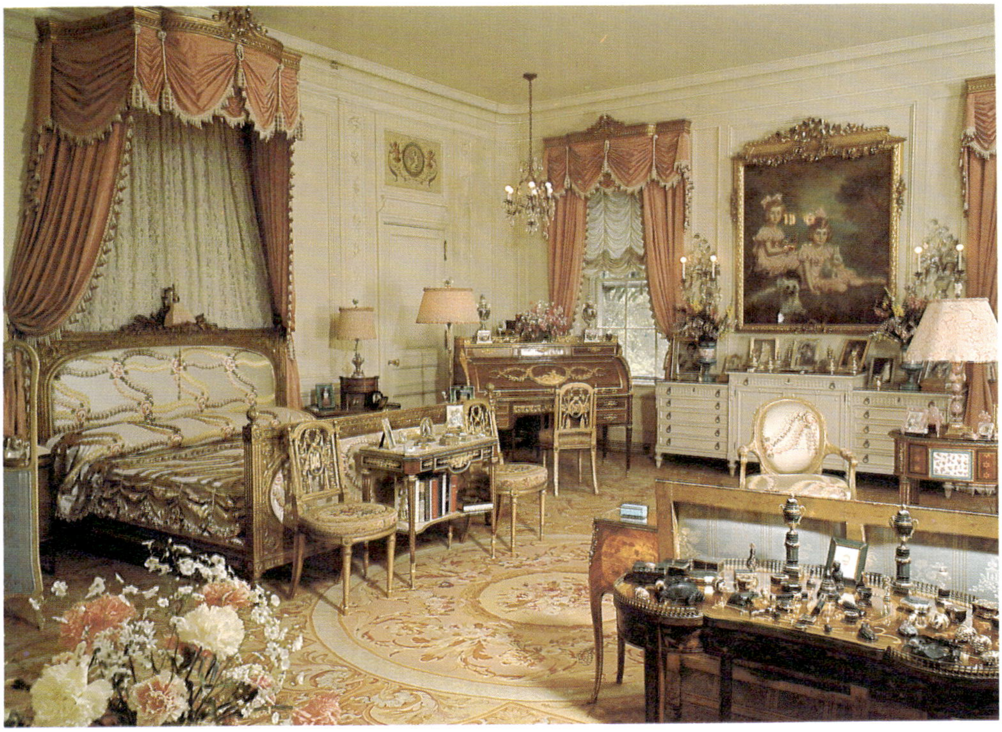

Despite an apparent slight yellow cast, this reproduction of an interior scene photographed in color conveys a good impression of the room's elaborate furnishings. (Compare this with the same halftone on the book jacket.)

aration called the *yellow printer*. (Yellow, it will be recalled, is the complementary color of light combining both red and green, by using yellow ink to reproduce the yellow printer's image, the original's red/green content is recorded on paper.) A similar procedure yields a *magenta printer* after the original is photographed through the green filter. The red filter produces the *cyan printer*, "cyan" being the printers' name for the cold "true" blue that is the mixture of green and violet-blue in the spectrum. However, magenta and cyan are often also called *process red* and *process blue;* this can prove misleading to people who confuse these shades with the red and blue of the light primaries.

One look at the pigment and light primary diagrams shows why the so-called process colors have to be yellow, magenta, and

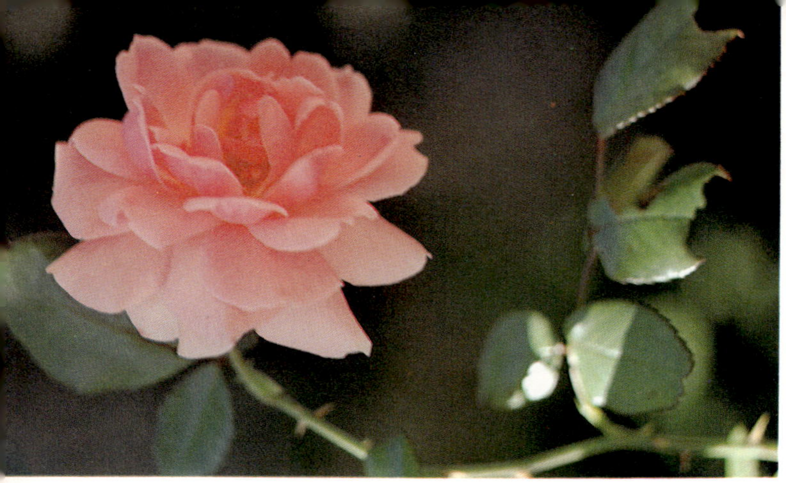

The diagrams on the right show the individual steps of photomechanical color reproduction. By the use of four filters, the original color photograph on the left is separated into the three primary pigment colors plus black. The resulting single-color images are shown on the far right.

cyan if the original picture's full color range is to be realized in the printed reproduction. As complementary colors of light, they contain—or rather reflect—two light primaries each, and when properly combined, produce all the other hues of the spectrum. Magenta and yellow yield the entire range of orange and warm red shades; yellow and cyan, all the greens from light to dark; and magenta plus cyan, the various violets and purples. Mixtures of all three result in a variety of browns that shade into black where they are combined in full strength. By utilizing the subtractive qualities of pigment color mixtures, every shade from light to deep and from bright to subtle can be re-created in the printed image.

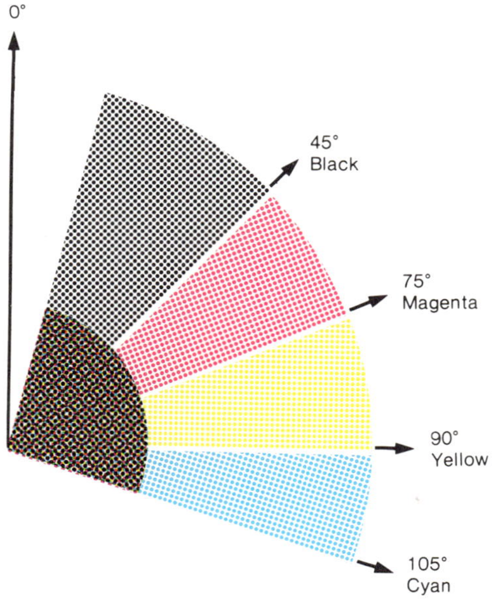

The different angles at which the four screens of a process halftone have to be tilted to avoid moiré patterns are demonstrated in this magnified section of a four-color print.

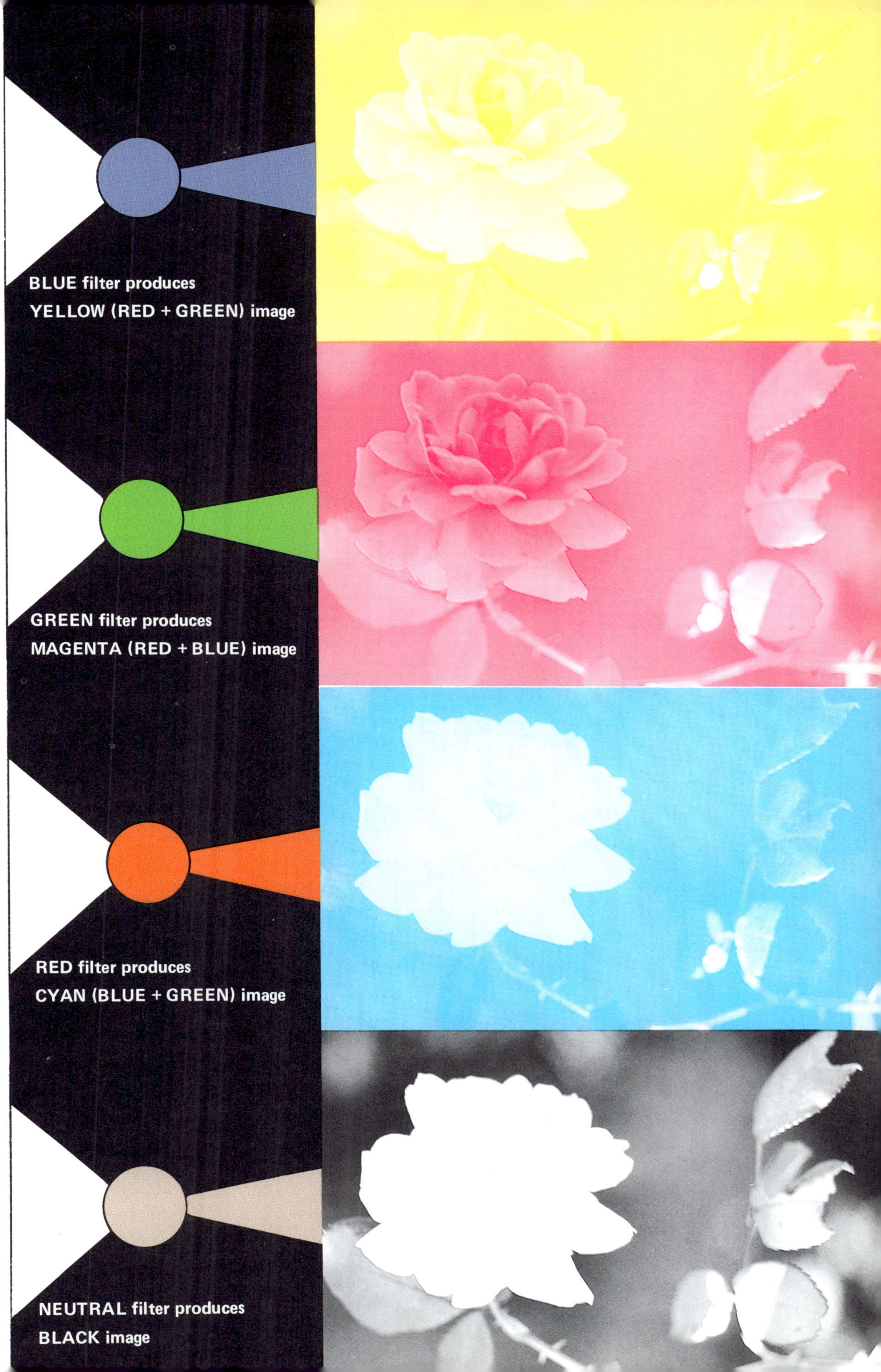
BLUE filter produces
YELLOW (RED + GREEN) image
GREEN filter produces
MAGENTA (RED + BLUE) image
RED filter produces
CYAN (BLUE + GREEN) image
NEUTRAL filter produces
BLACK image

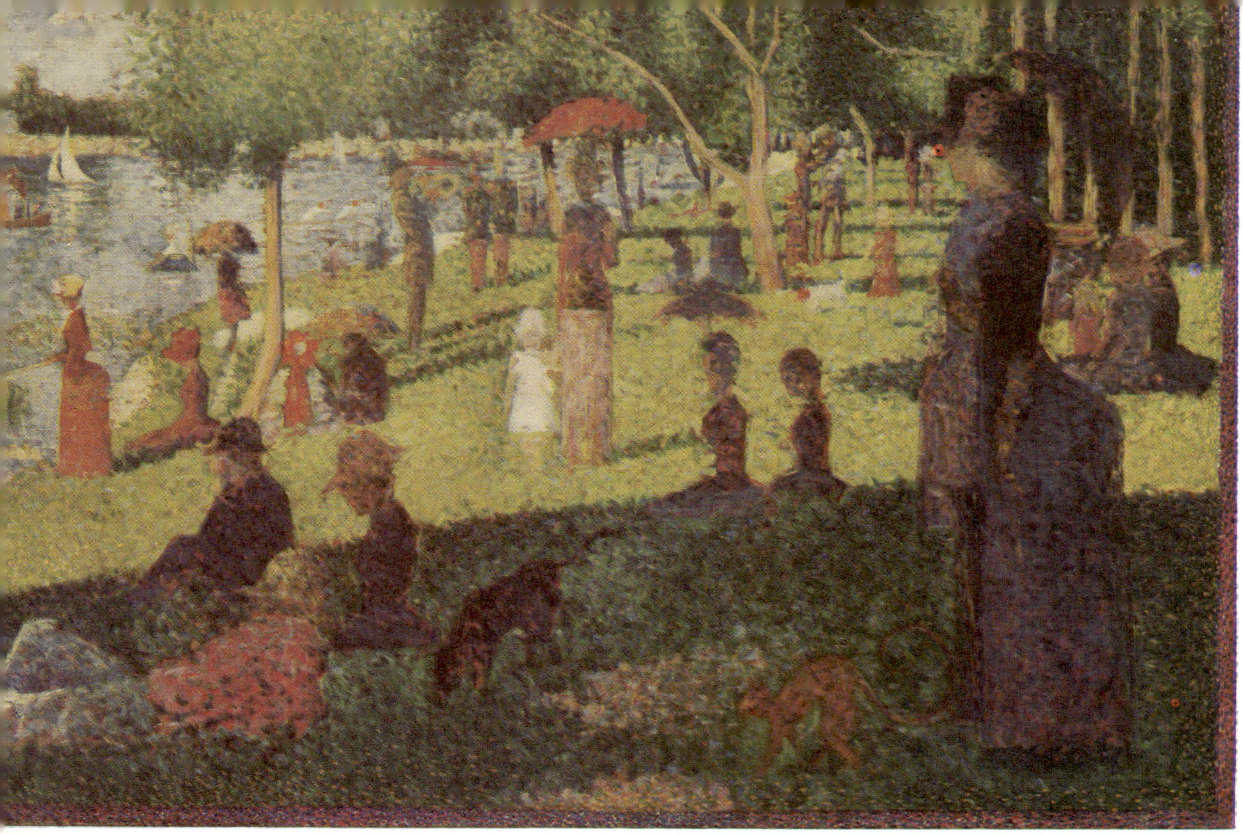

Process reproduction of an oil painting by Seurat. Only a comparison with the original can establish the degree to which any reproduction maintains fidelity of color values.

That, at least, is the theory. According to that theory, the three pigment primaries in their various combinations should suffice to reproduce the original in all particulars. Unfortunately, however, it does not happen quite like that in real life because machines are not sensitive enough, and the color values of the printing inks not pure enough. Hence color corrections usually have to be made. There may be too much of one and too little of another color, resulting either in color loss or a "red cast" or a "blue cast" or some other color defect in the printed image. In addition, the shadows appear brownish and the illustration looks flat, so that definition as well as three-dimensional effects are missing. In order to overcome some of these defects a fourth photograph taken through a neutral filter is made for the black plate. Overprinting this black impression on the three combined colors invests the final image with definition, depth, and body.

Obviously the photomechanical process of color separation is complicated and depends on a variety of factors for success. One problem not encountered in screenless printing methods is the danger of producing moiré patterns when the various color screens are superimposed on one another. In order to avoid such patterns, the printer angles the screen differently for each color. The margin of error here is very small; screens that are "off" by no more than one-tenth of one degree, or are slightly maladjusted because of defects in the register, may produce undesirable patterns in those areas where all the colors print.

Perfect register—the exact correspondence of position in the four imprints that make up the full-color reproduction—is important for other reasons. Even when moiré patterns are not a problem, imprints that are off register even slightly may destroy fine detail and upset the color balance in the reproduction.

The use of electronic and computer equipment has reduced substantially some of the difficulties inherent in process separation. Electronic scanners, for example, can determine the color values of the original with the help of filtered photocells, which then feed the information into a computer. The data from the computer show exactly how and where the color values of the original differ from those of the separations.

Despite this advanced technology, some difficulties remain because so many factors ranging from the quality of the camera work and the printing to that of the paper and the inks employed play a part in determining the degree of success achieved in reproducing the original. Good camera work is, of course, critical, and here it is not primarily the more or less standard equipment and materials that count, but rather the human element, meaning the camera and scanning expert's training, experience, and almost instinctive knowledge of how to best handle each individual subject. Today, most color transparency separations are made by electronic scanning devices, but original artwork that is either too large or prepared on too rigid a material for mounting on such scanners still requires the use of conventional camera techniques.

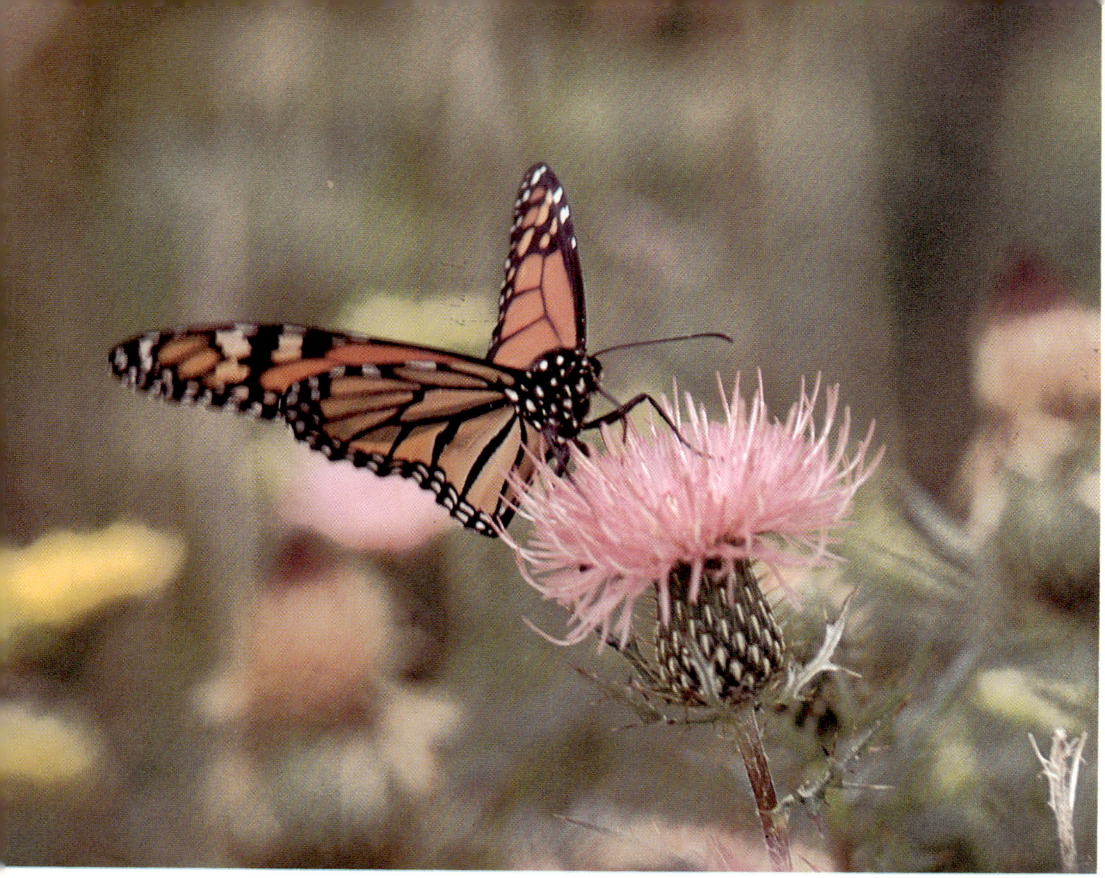

Process reproduction of a photograph focusing on a monarch butterfly exploring a thistle blossom for nectar.

Once the book is on press one of the main concerns aside from perfect register is the maintenance of even ink coverage over the large sheet. Although in theory such coverage is assured, in practice variations often do occur, and these may become a problem mainly because of the *imposition,* the relative position of the pages on the sheet that will produce the correct page sequence in the book after the sheet has been folded for binding. This means that pages facing each other in the book are not so located on the flat sheet except where the imposition of the book permits it for the two facing pages that make up the center of each *signature,* or binding unit, which in most instances comprises either sixteen or thirty-two pages. For the rest, facing pages may be widely separated, so that some double-spread illustrations may print in two halves located on pages at opposite ends of the sheet. Any appreciable difference in ink coverage at one end will become disagreeably obvious after the two separated halves of such illustrations are joined together in the bound book. In order to obviate that possibility as well as the binding difficulty of making the two split halves meet accurately in the center of the fold, it is preferable to plan double-spread illustrations only for the center spreads of each signature, where they will be free of problems. If such an arrangement is not feasible, at least the most sensitive illustrations should be so located.

The great divergence in the quality of color printing is evident to anyone who sees widely varying reproductions of famous masterpieces, and has had the chance to compare a printed image with the original painting on display in a gallery or a museum. Some prints are so badly "off color" that they are hardly recognizable, whereas others are remarkably faithful both in color and in fineness of detail.

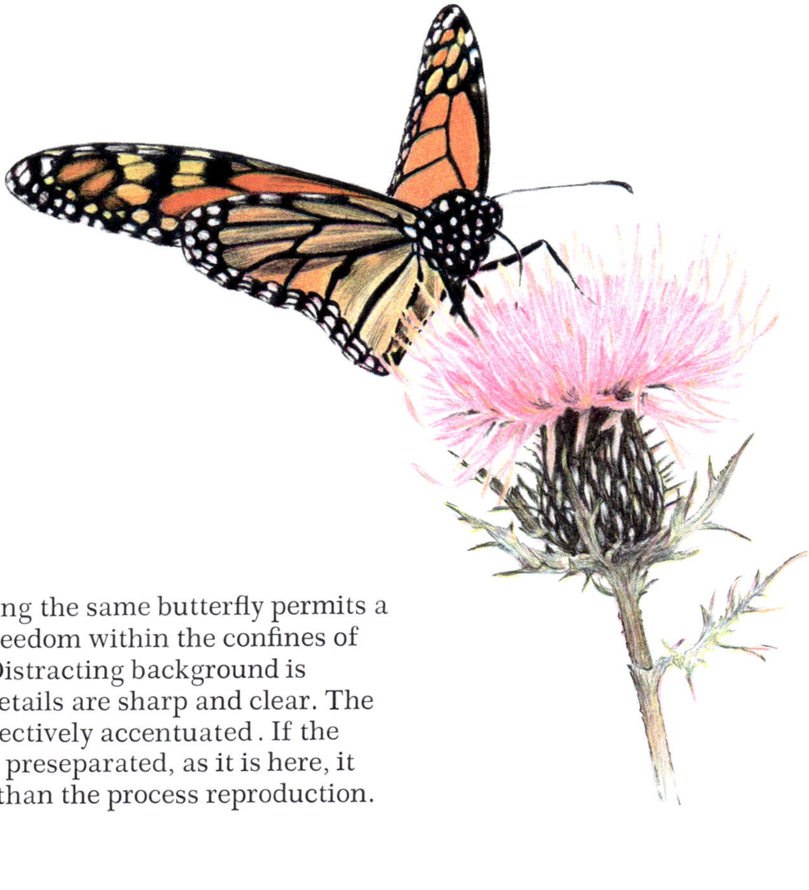

Hand-rendering the same butterfly permits a measure of freedom within the confines of the subject. Distracting background is eliminated; details are sharp and clear. The colors are selectively accentuated. If the illustration is preseparated, as it is here, it will cost less than the process reproduction.

The paper used for book printing represents a large part of the over-all reproduction cost and also has a great bearing on its outcome. Paper quality and characteristics influence the appearance of the final product and especially the fidelity of the colors. The best choice for screened halftones, and an almost indispensable one for delicate process color work, is a coated paper with a smooth and firm matte or glossy surface. This kind of paper permits the reproduction of very fine halftone screens with sharp definition and exact color values. A softer, less smooth paper does not reproduce fine screens as well; it also allows the colors to spread and "bleed" to a certain extent, thereby changing the tint strengths. However, many high-grade noncoated papers designed for offset printing are perfectly adaptable to all except the most sensitive color work, and yield satisfactory reproduction of most line and halftone illustrations. An artist planning full-color illustrations should inform himself in advance about the kind of paper that will be used in printing them, for its characteristics may affect the color values in the reproduction and therefore the degree to which the artwork should compensate for such variations.

The relationship between the properties of paper and those of inks is another important factor; some papers demand specially matched inks for successful printing. By and large, however, the standard offset inks are designed to print well on most papers used in book manufacturing. Offset inks are distinguished by their very strong color values, a fact that compensates for the relatively light

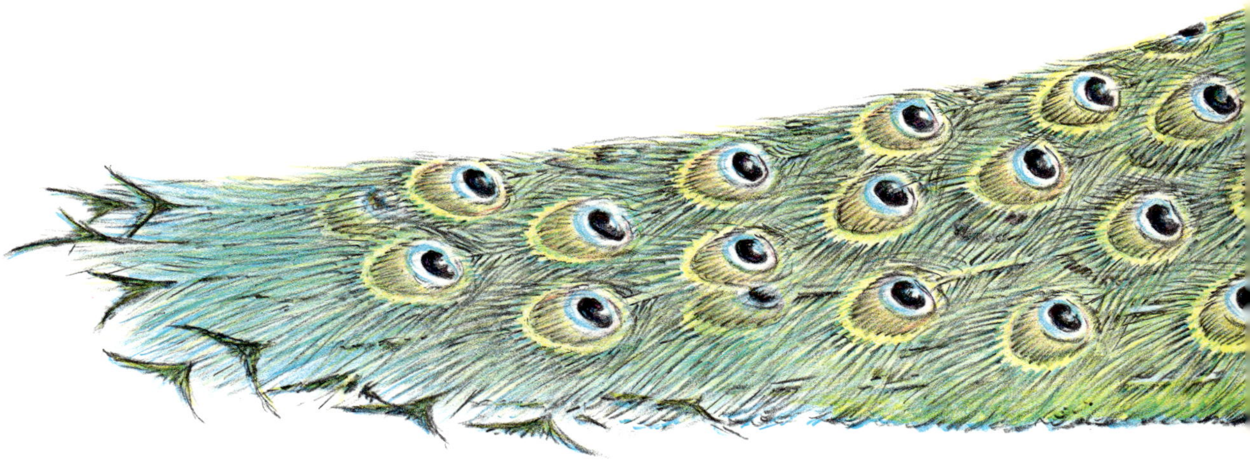

ink application that is characteristic of offset lithography as compared to other printing methods.

Despite the many labor-saving improvements of the modern printing process, good photomechanical reproductions of original full-color illustrations, be they photographs or hand-rendered artwork, require a considerable amount of time and effort and are therefore expensive. Understandably, therefore, publishers welcome alternatives for books that require full-color illustrations. Preseparated artwork, in many ways an old art that has been rediscovered and adapted, offers the great advantage of dispensing with the need for the long and costly process of separation by photographic filters and the subsequent chore of color correction. It is therefore an attractive alternative to the use of full-color originals.

This illustration of a peacock is a good example of the purity and brilliance of color that may be achieved by screenless preseparation.

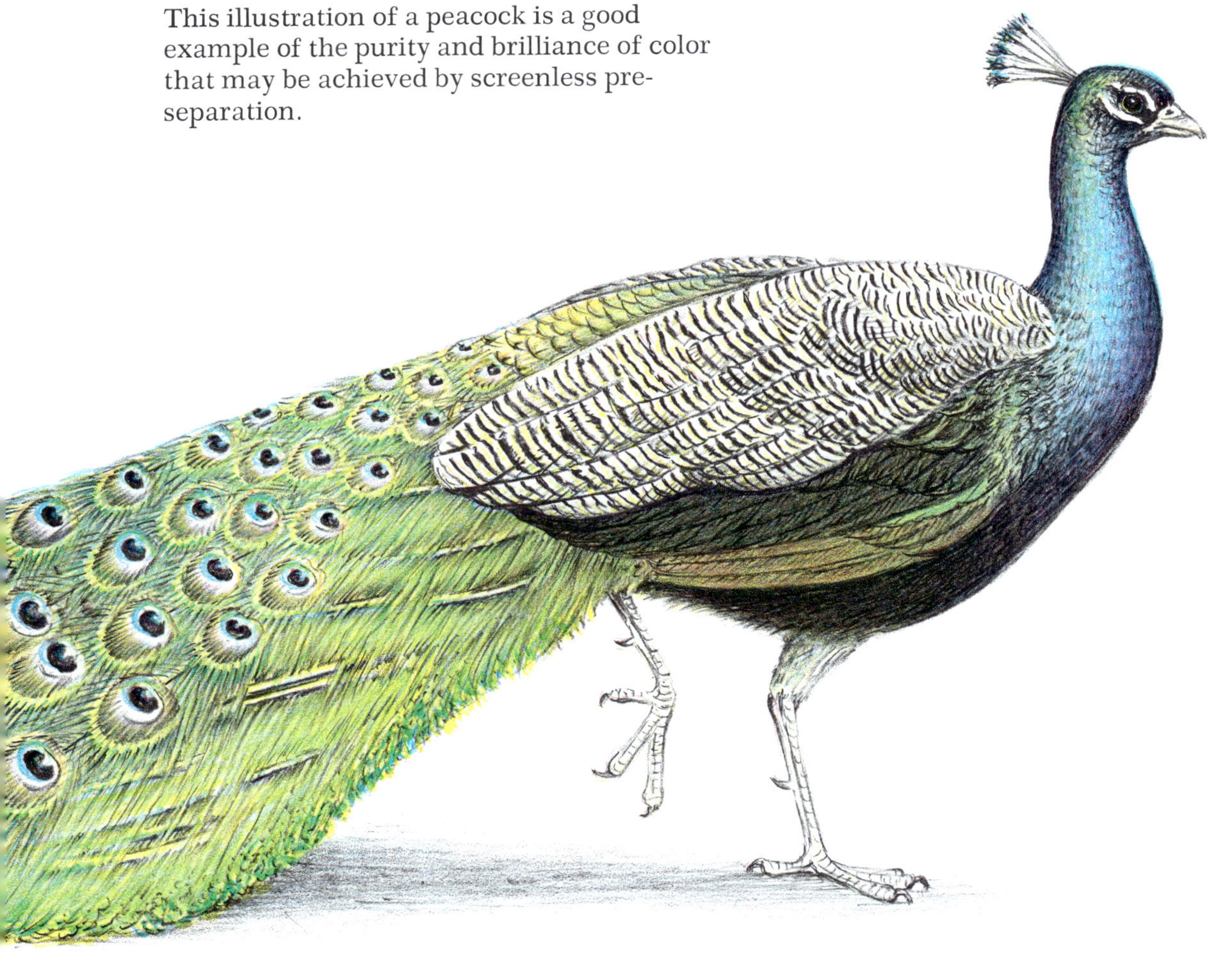

For the illustrator, preseparated four-color artwork is an ambitious and challenging undertaking. Creating a perfectly balanced set of four black separations, a set that will produce a result comparable to the "process" image of a full-color original when printed in the primary colors and combined, is obviously not a simple task. Such artwork involves a great deal more than adding one or more attractive colors to a black drawing in order to "dress it up" and make it look more lively and appealing. This in no way implies a judgment of quality; preseparated illustrations executed in four nonprimary hues chosen for their ornamental value only may be extremely handsome and the perfect complement to texts such as fairy tales. They also have the great advantage, at least from the viewpoint of the artist, that deviations from the colors originally planned will hardly be noticeable, and that none of the combinations need produce any specific, narrowly defined hues designed to match those of an animate or inanimate model.

Wherever preseparated artwork is intended for full-color illustration designed to reproduce faithfully the image of a bird, flower, or seashell, or other natural or man-made object, it becomes a different matter entirely. Gone is the artist's freedom to exercise his imagination and choose almost any appealing hue, applying the color in ways deemed ornamental and attractive. Instead, the artist's eye will in effect have to perform the work of the four camera filters in separating the various hues into their component colors and transferring these values, accurately translated into black percentages, onto the four separations.

The magnitude of the task can be best demonstrated by visualizing an actual example such as a book on wild flowers requiring some eighty full-color illustrations. Printing the book in black only would substantially reduce its value; on the other hand, process separations of full-color originals—photographs or watercolors, for instance—would increase the production costs beyond acceptable levels. Preseparated artwork for such a book is an attractive alternative, provided the artist can by that method create illustrations of high quality and at the same time faithfully portray the various shades and hues of blossoms and foliage that distinguish each species. In order to be successful, the artist's eye and mind may have to separate and identify several thousand different color combinations.

So formidable a task can be managed well only by illustrators who possess several special skills and abilities. Presupposing basic draftsmanship, they also have to have a feeling for the balanced arrangement, or composition, of pictorial ingredients, a highly sensitive eye for color distinction, and the skill to translate values so perceived into black percentages. In addition, such illustrators must be well acquainted with the mechanical aspects of color reproduction, for without this technical knowledge, they will not be able to appreciate the possibilities as well as the limitations of the printing process by which their artwork will eventually be reproduced.

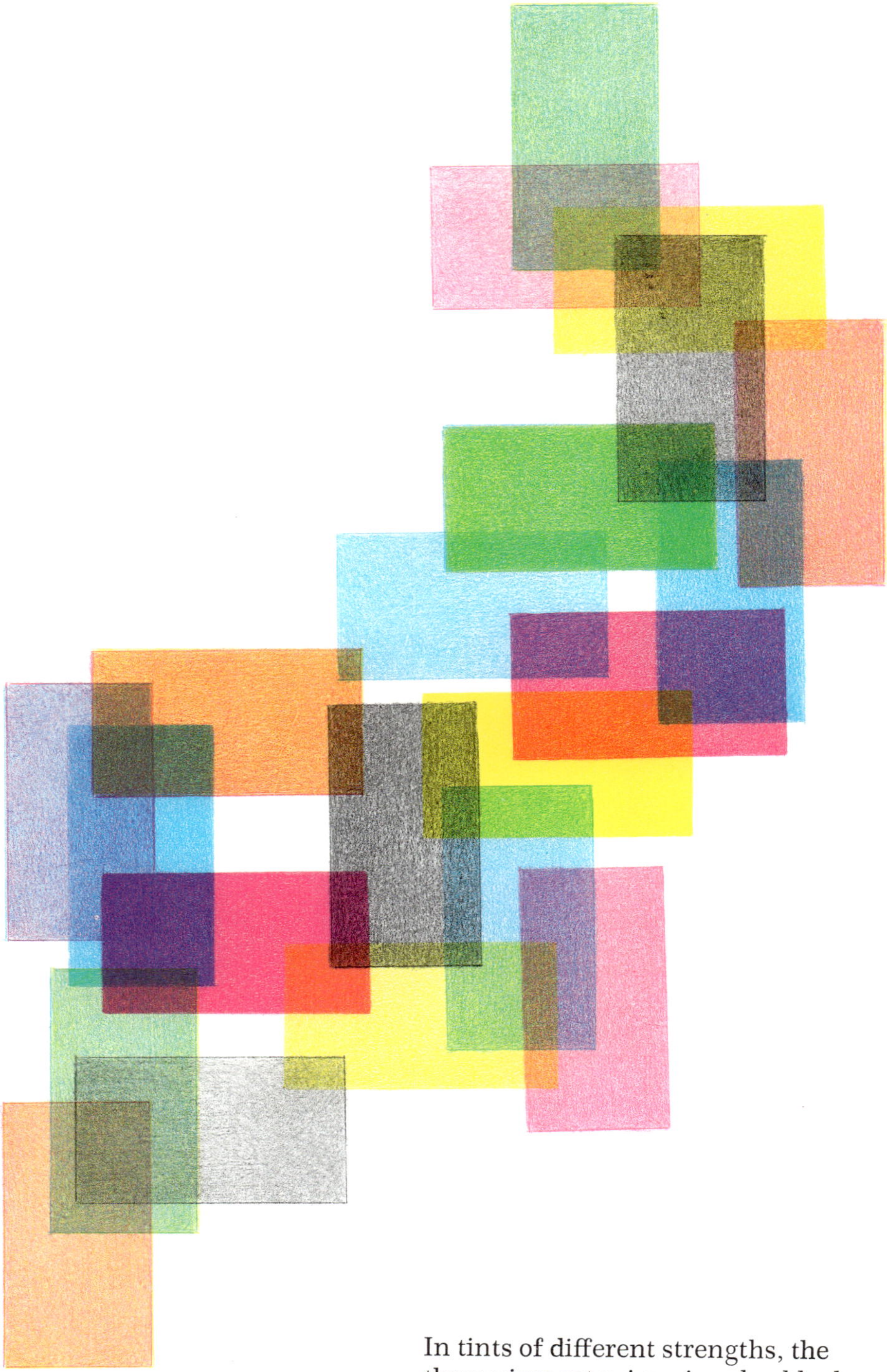

In tints of different strengths, the three pigment primaries plus black yield a great variety of combinations.

The Infinite Variety: Color Combinations

That most colors in our environment, especially in nature, consist of mixed hues is a fact appreciated by everyone who has tried to capture these colors on canvas or paper. Even a small patch of plant growth will serve to demonstrate a great variety of subtle greens and browns. It has been said that seventeen different shades of green may be counted in a famous watercolor by Dürer of a small chunk of soil lifted from a meadow with its microcosm of mosses, grasses, and flowering plants.

Even bright, pure-looking hues such as those of many birds, flowers, and butterflies turn out to be combination colors upon closer examination. There will be reds with various admixtures of blue or yellow; yellow tinged with green or warmed by reddish and brownish tints; blues that glint with unexpected greenish or purplish highlights, and purples deepened by blue-black shadows. Added to these is the seemingly endless array of delicate "in between" hues of pink and lilac and orange and chartreuse. The result is a wide range of color combinations that are as attractive to the beholder as they are challenging to the artist. Rarely will any one of these shades reflect just a single-pigment primary color; most are mixtures of at least two and more often all three primaries.

To those involved in printed color reproduction that last fact is of more than merely academic interest, for it is the balance or imbalance of colors in a printed image that makes the difference between a good and a poor reproduction. Poor work suffers from either too little or too much of at least one of the combination colors; such variations, even slight ones, in the different values of one primary affect all the combination hues and therefore the entire reproduction. The printed sheet may, for example, have a "bluish cast," meaning that the blue imprint is too strong. This will distort not only the various blues in the illustration but also the greens, browns, and purples. In addition, any increase in the blue content of a picture makes the reds and yellows appear "warmer" than they should compared to the other colors. The opposite, of course, is true of any loss of strength. Such "difference-by-contrast" optical color illusions are an important factor in almost every kind of color reproduction; used deliberately, they can be very helpful in creating specific effects.

Optical illusions are more often associated with shapes and lines than with colors. Whatever they involve, all optical illusions are alike in that they fool the observer's eye into seeing something as being different from what it is in reality—perceiving a straight line as bent, for instance, or geometric shapes as distorted. Although few people are aware of it, optical illusions involving colors

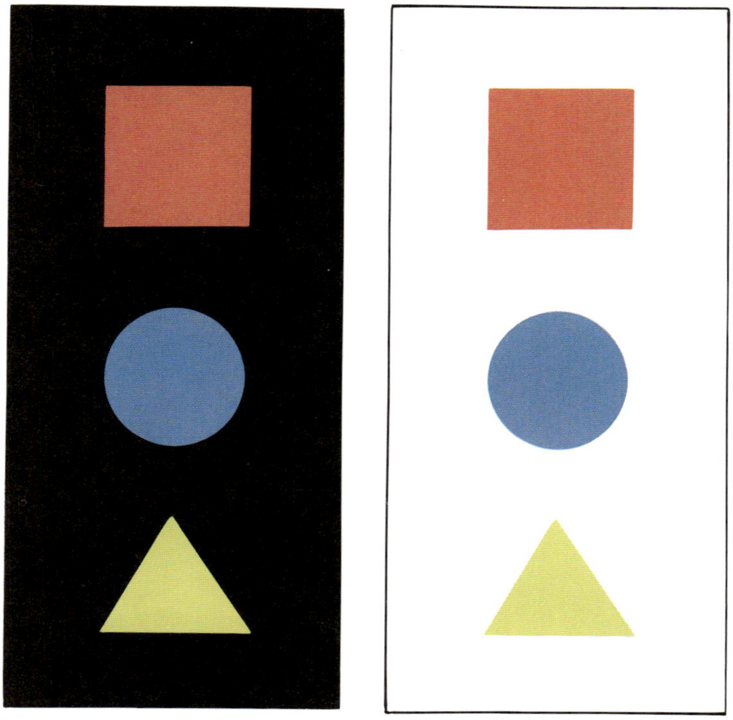

In one of the classic examples of how strongly colors are affected by their surroundings, the red, blue, and yellow shapes on the black background appear much lighter than the same colors printed on white.

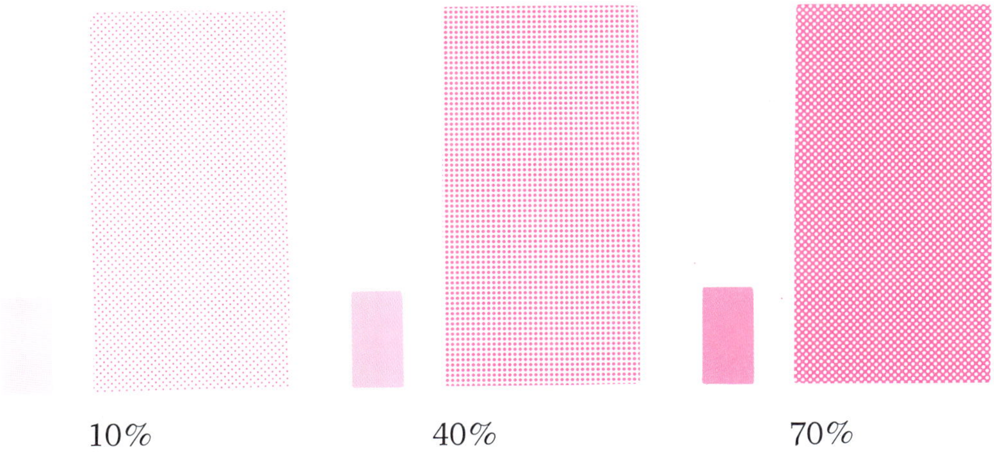

An enlargement of red tint strengths in percentages of 10, 40, and 70 show exactly how the illusion of pink is achieved by printing dots of solid process red on a white background. The strength of the tint is determined by the relationship between the size of the dot and that of the surrounding interspaces. The eye mixes the solid color and the white for the appropriate color gradation.

are part of our everyday visual experience.

A good example is the "halftone" process of color printing, which uses screens to break up areas that have to reproduce as less than 100%, or solid, color. Although a screen tint of red looks like pink, that pink is an optical illusion. The tiny dots that make up the screen pattern are printed in solid, 100% red, but they are very small and in the lighter tints surrounded by white spaces that are much larger, which results in the observer's eye mixing the red and white to produce pink in his mind. Any decrease in the dot size—and corresponding increase in the amount of white interspace—produces a paler pink. As the size of the dots increases and that of the white spaces decreases proportionately in the higher percentage tints, the pink becomes deeper. The reverse pattern—white dots on a red background—is used for the deepest tints from 60% onward, until at 90% the tiny white dots resemble in size the equally tiny red dots in the 10% tint.

The fact that colors are affected by their neighbors in many ways was always appreciated by painters, but in the nineteenth century

French painters of a school that was to become known as *impressionist* made a special study of this phenomenon. They were greatly influenced by the work of Michel Chevreul, who was a chemist rather than an artist. However, as director of the famed Gobelins rug factory of Paris, Chevreul was intimately involved with color, and complaints about the quality of some of the Gobelin colors led him to study the effects produced by contrasts between the various chromatic hues and their effect upon one another. This study became a lifelong interest and preoccupation with Chevreul.

French impressionists patterned much of their technique along lines suggested by Chevreul's work; employing only the colors of the spectrum and striving to use the various pigments in as pure a form as possible, they banished the brownish "earth" colors such as umbers and ochres from their palette. So obsessed did some of them become with the goal of creating pure light effects by manipulating the pigments with which they had to work that they formed a special division of the impressionist school called *luminism,* or "light painting." A related group known as the *pointillists* employed the technique of placing small daubs, or dots, of pure color side by side and letting the eye of the beholder do the mixing. They believed that a large mass of tiny individual yellow and blue dots, for example, could create a green much purer than any shade mixed from blue and yellow on the palette.

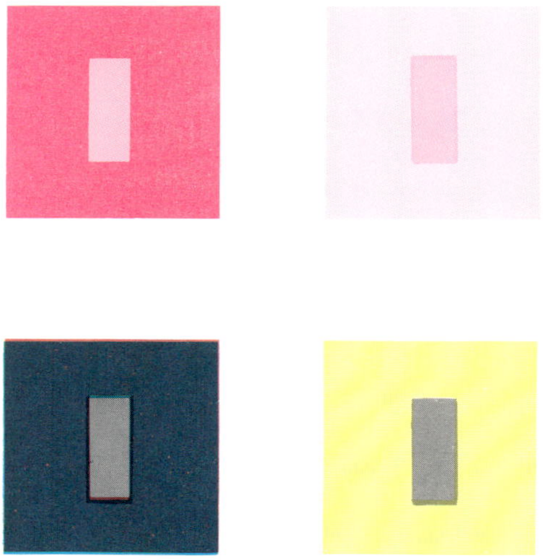

Even though they may look different, both small tint rectangles inside each of the two sets of color squares on the left are identical and correspond exactly to the tint of the larger rectangle on the right. Very deep or very light colors influence neighboring hues to the point of visual illusion.

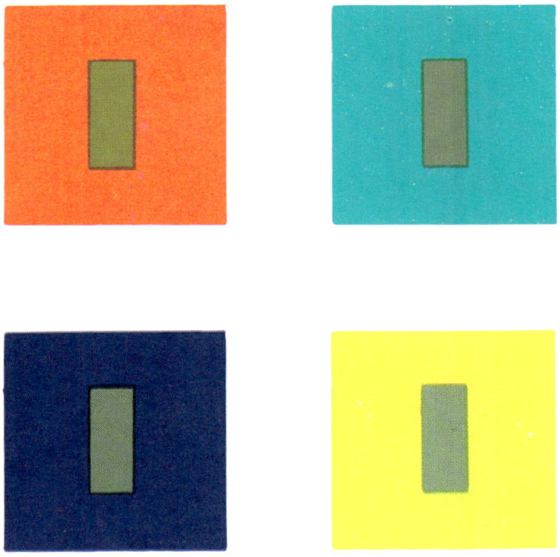

In this example, the colors selected for the squares on the left create the illusion that both rectangular enclosures are similar when in fact they are different, as shown by the larger rectangles on the right. "Cold" and "warm" colors profoundly affect the appearance of neighboring hues.

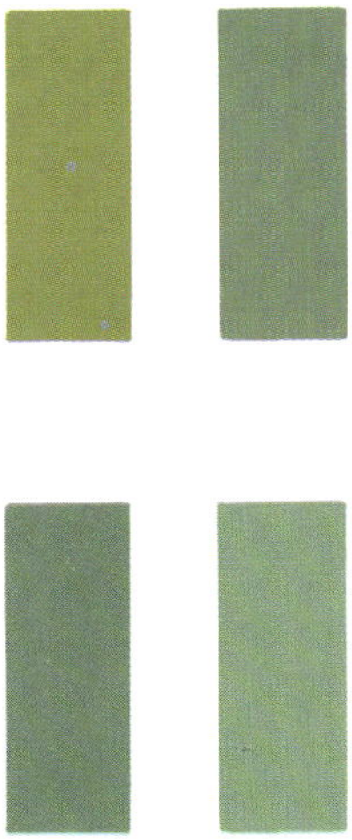

Although artists working with pigments are fully aware that color strongly influences neighboring shades, this fact often surprises those whose experience with color is limited. That accounts for the frequent disappointments of people who select specific hues from a printer's chart or swatch book where small color squares are neatly lined up in rows, each separated from the other and surrounded by white paper. Later, when these people see their selection printed in larger areas, and then perhaps not bordered by white but instead in combination with other colors, and perhaps reproduced on a blue-white or off-white paper, they can hardly believe that the ink on the final print represents the hue of their original choice.

When a tint block of 40% red is placed on a larger square of only 10% red, the enclosure may look like a fairly strong pink, yet the same tint will appear pale on a similar square of 100% red. If the two squares are placed side by side, most people take the two enclosures for different shades of pink even though they are in fact identical. A pale color enhances, and a strong color weakens, any hue it surrounds.

If certain chromatic constellations can make identical hues look different, the reverse is also true: different shades may appear identical because of neighboring colors. All of this is very important to anyone concerned with color reproduction, especially to artists

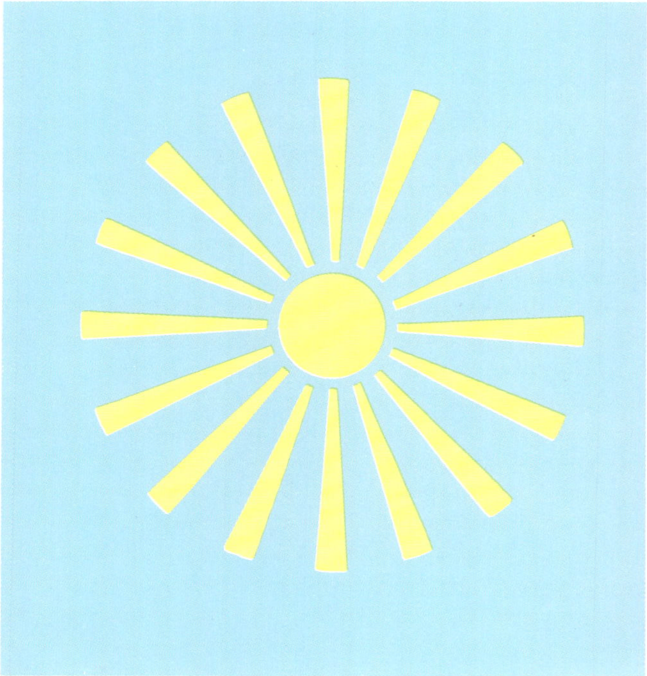 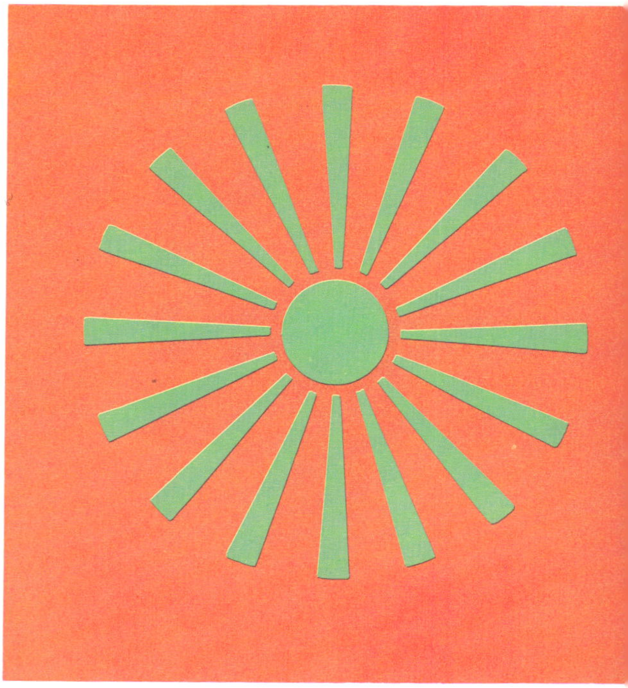

engaged in preseparations. Because they must work "blind," as it were, in a black medium with gray values, and cannot see the effect of their color combinations until the illustrations are printed, they have to visualize every shade and compensate for possible optical illusions. As a concrete example, let us assume an illustration of red and blue flowers, both of which are to have identical-looking orange-yellow centers. In order to make these centers look the same, the one in the blue flower has to get less red because the surrounding blue accentuates and emphasizes the warm color.

Another problem in chromatic juxtapositions is that of color clashes. The unpleasantly shimmering effect of certain color arrangements is a familiar phenomenon; it affects some people so strongly that they feel dizzy looking at such colors. What actually happens is that the receptors in the eye are irritated and fatigued by the vibrations caused by the immediate proximity of certain hues. Miraculously, however, the same unpleasant "clashing" colors appear perfectly acceptable to the eye if they are separated by black

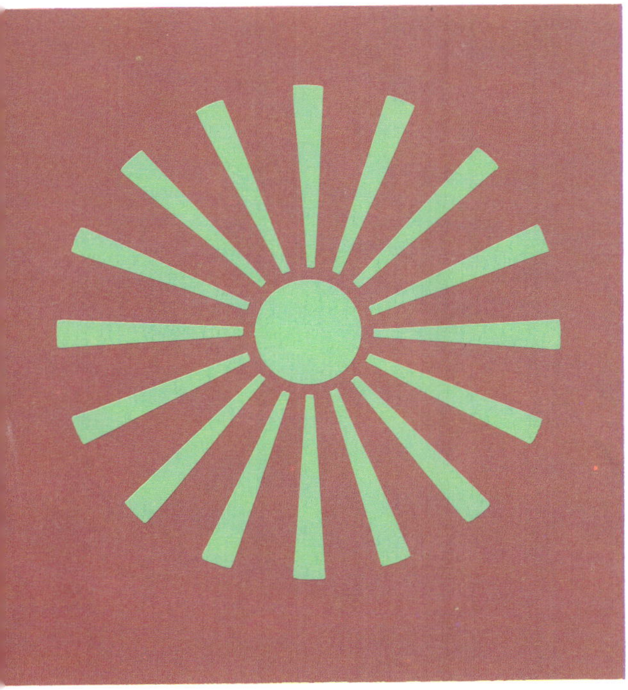

Printed in yellow on a blue background, the sunburst design is visually pleasant in the first illustration, but causes unpleasant shimmering vibrations when colors are changed to green and red in the second picture. By adding a blue tint to the red, the design is once again rendered visually acceptable in the third illustration.

lines. This principle has been used for ages by designers of stained-glass windows, who also knew that outlining them in black makes colors appear more vivid; even pale tints that normally look very similar take on a character of their own when enclosed by black lines. These facts have been utilized not only by creators of stained-glass objects, but also by artists employing the technique known as *cloisonnisme,* or compartmentalization; among the modern painters, Matisse availed himself frequently of this method. One reason for the fact that colors so rarely clash in nature is that the shadows cast by three-dimensional objects form their own dark borders separating them from adjoining color areas.

Any artist engaged in preseparations has to be aware of all the factors capable of influencing his color effects in the eventual printed image. The greatest challenge, however, is the accurate evaluation of the almost unlimited variety of color combinations and their translation into black percentages. Any assistance in determining the exact values of different hues and their compo-

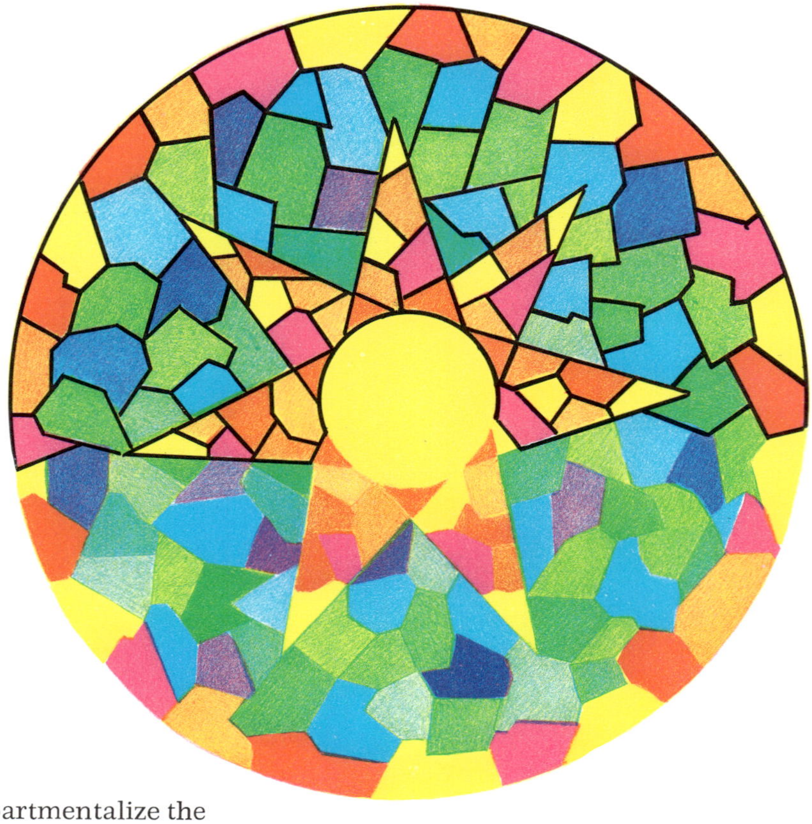

Black lines compartmentalize the color blocks in the upper half of this mosaic and so create a stained-glass effect which makes the enclosed hues appear more vivid than those in the lower half.

nents by means of sample charts is of great importance, and reduces the likelihood of misinterpretations.

Color charts showing process color combinations in various tint strengths come in many different shapes and sizes. All of them are to some extent helpful, but most of them do not present the material so that it is easy or convenient for an artist to use. There are many repetitions, the sequence and color content are not always clear, and the great number of little color squares, while visually attractive as a pattern, is often also quite confusing. But then these charts are made up by printers and ink companies for their own use rather than for that of an artist who needs to analyze and mentally separate the various color combinations he is supposed to render.

The color charts reproduced in this book are designed to fit the needs of an illustrator. Based upon many years of experience with full-color preseparated artwork, they concentrate on the essentials and combine several advantages. The same key diagram is used throughout; shown in black on the first page of charts, it is made up of a block of four different tints in values of 10%, 30%, 50%, and 70% grouped around the 100%, or solid, block in the center. In this way, each tint is bordered by white paper on the outside and by other shades on the inside.

The charts on the first two pages feature two-color combinations. Diagram blocks of the primary colors shown at the far left are repeated five times in each row, each time with the addition of a different tint of a second primary color. The result is a total of twenty-five different shades for each of the three rows of two-color combinations.

On the following double spread the third color is added for a total of one hundred different three-color combinations. Since these multicolored hues are usually the most difficult to analyze, and since they make up most olive greens and browns, such extensive coverage is highly desirable.

The last two pages of charts show various percentages of black added to one-, two-, and three-color diagram blocks. These mixtures demonstrate very clearly the deadening effect of black when added to chromatic colors; although this effect may be very useful and even necessary in many instances, there is always the danger of overstating the black in an illustration. By and large, it is best to use black carefully and with a light hand so that it will be a supportive, strengthening factor in the final reproduction instead of a deadening one.

Those who learn to employ properly the great variety of pigment color combinations that span the scale from warm to cold and from delicate to brilliant can produce very individual artistic results in illustrating.

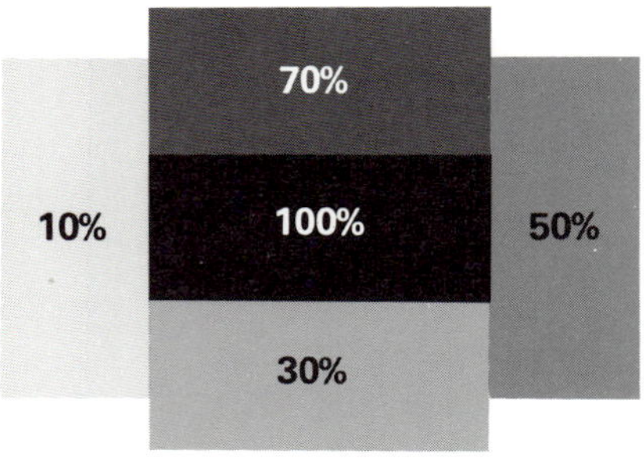

Key BLACK tints

The block of four key tints plus solid color in the center, shown in black on the left and repeated below in the three primary colors, serves as a basis for all the color combinations on these and the next four pages. On pp. 84–85, four different strengths of blue added to the yellow-magenta combinations produce a wide

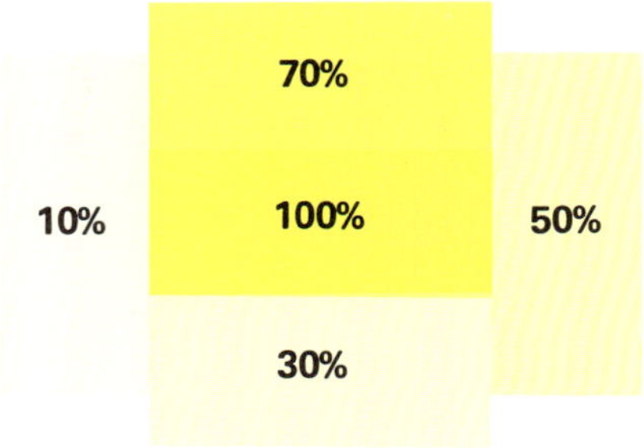

Key YELLOW tints

+10% PROCESS RED

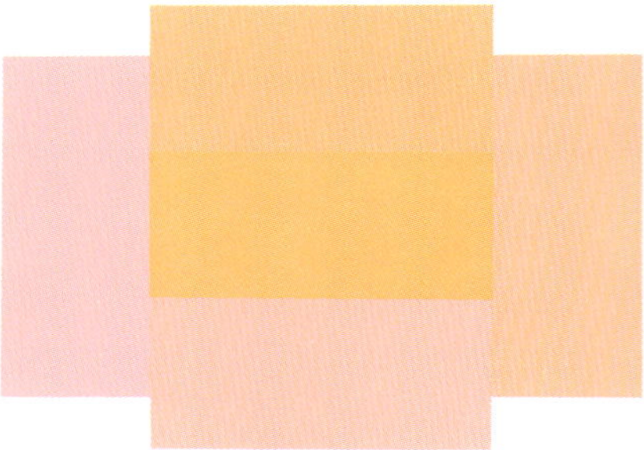

+30% PROCESS RED

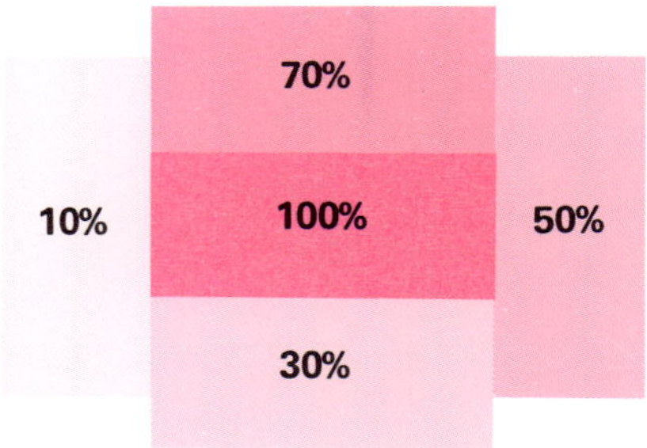

Key PROCESS RED tints

+10% PROCESS BLUE

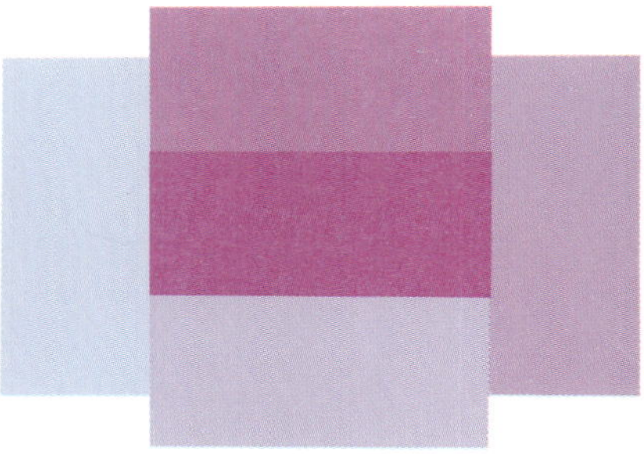

+30% PROCESS BLUE

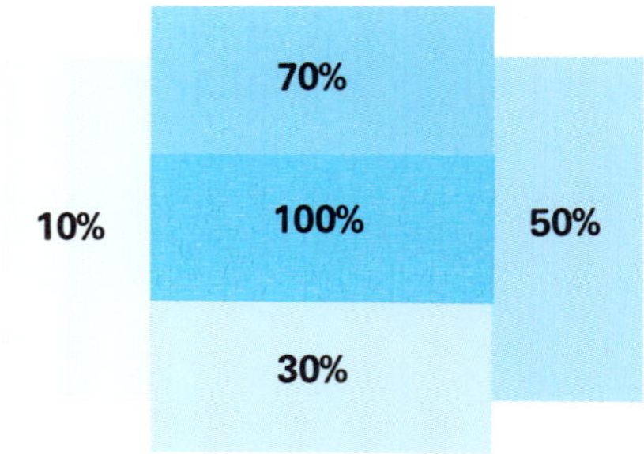

Key PROCESS BLUE tints

+10% YELLOW

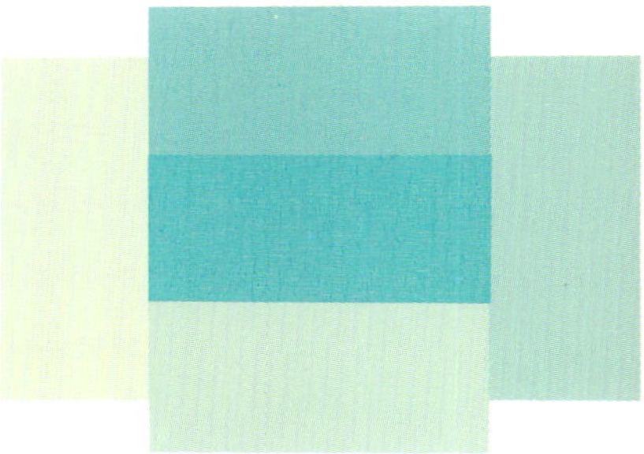

+30% YELLOW

variety of browns and olives in different shades. On pp. 86–87, black tints have been added to single primary color blocks as well as to two- and three-color mixtures. Altogether, over 300 different combinations are represented in these charts.

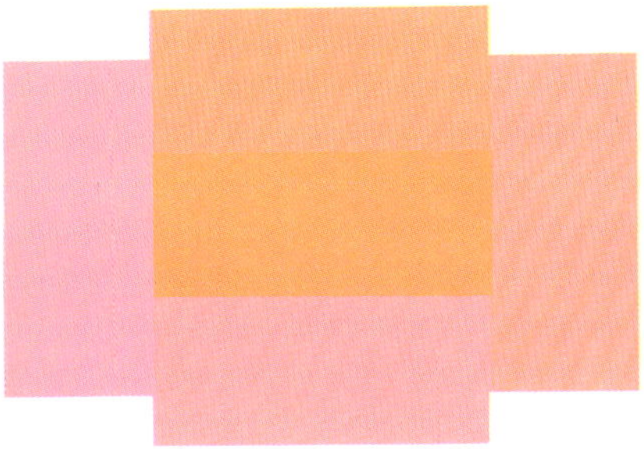

+50% PROCESS RED

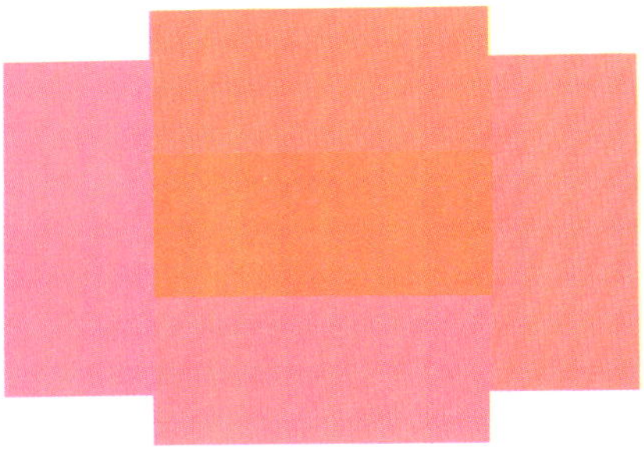

+70% PROCESS RED

+100% PROCESS RED

+50% PROCESS BLUE

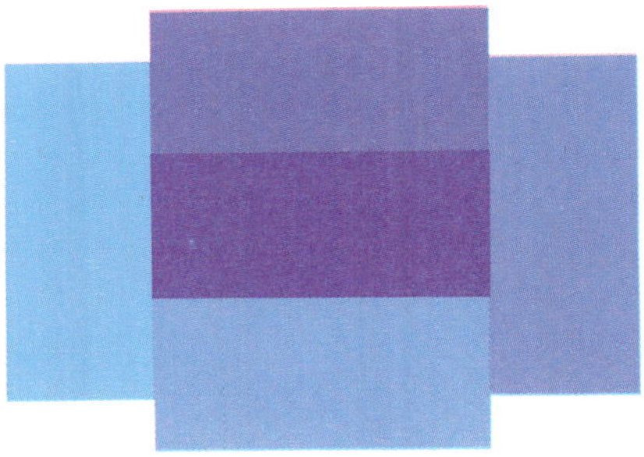

+70% PROCESS BLUE

+100% PROCESS BLUE

+50% YELLOW

+70% YELLOW

+100% YELLOW

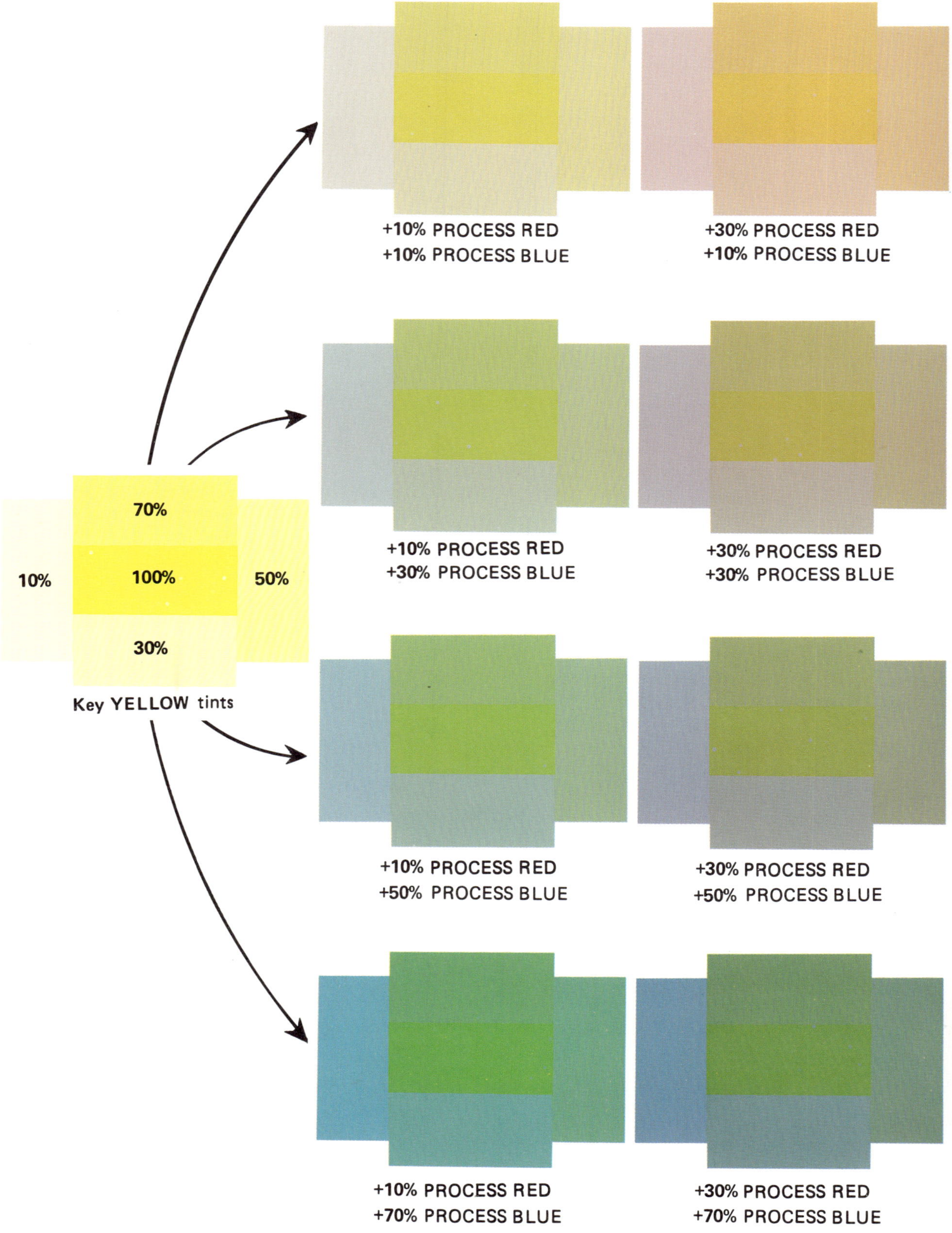
70%
10%
100%
50%
30%
Key YELLOW tints
+10% PROCESS RED
+10% PROCESS BLUE
+30% PROCESS RED
+10% PROCESS BLUE
+10% PROCESS RED
+30% PROCESS BLUE
+30% PROCESS RED
+30% PROCESS BLUE
+10% PROCESS RED
+50% PROCESS BLUE
+30% PROCESS RED
+50% PROCESS BLUE
+10% PROCESS RED
+70% PROCESS BLUE
+30% PROCESS RED
+70% PROCESS BLUE

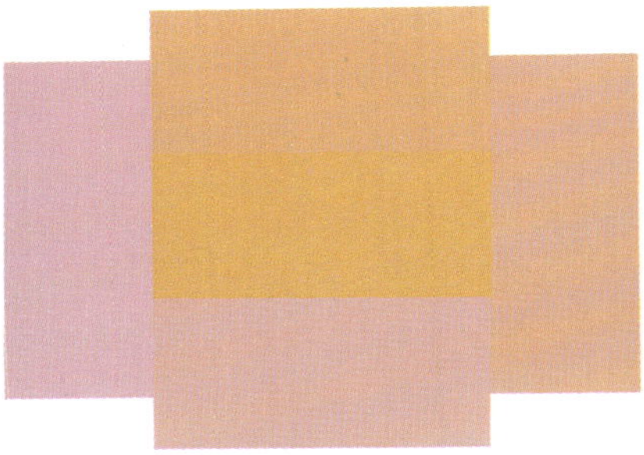

+50% PROCESS RED
+10% PROCESS BLUE

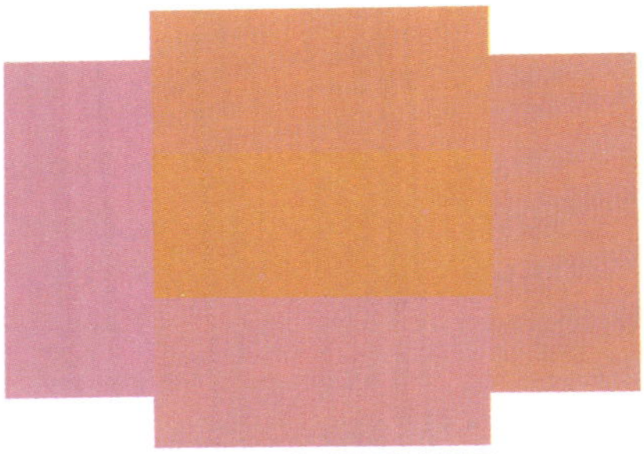

+70% PROCESS RED
+10% PROCESS BLUE

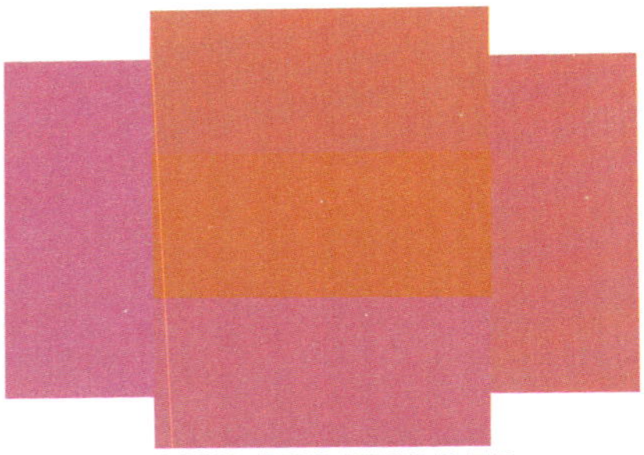

+100% PROCESS RED
+10% PROCESS BLUE

+50% PROCESS RED
+30% PROCESS BLUE

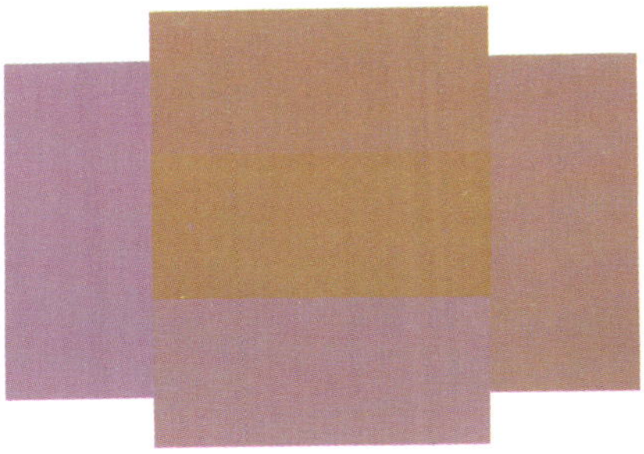

+70% PROCESS RED
+30% PROCESS BLUE

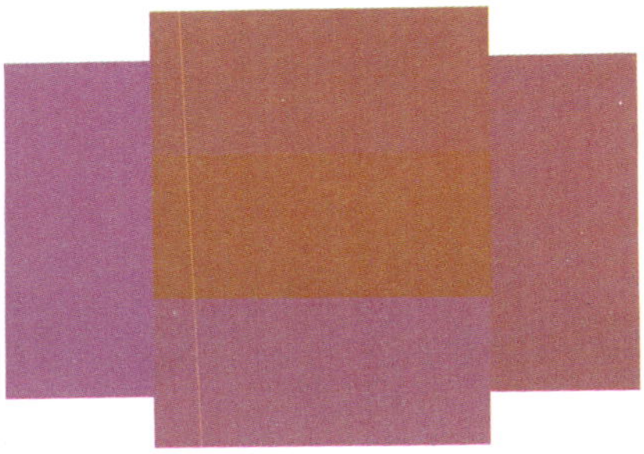

+100% PROCESS RED
+30% PROCESS BLUE

+50% PROCESS RED
+50% PROCESS BLUE

+70% PROCESS RED
+50% PROCESS BLUE

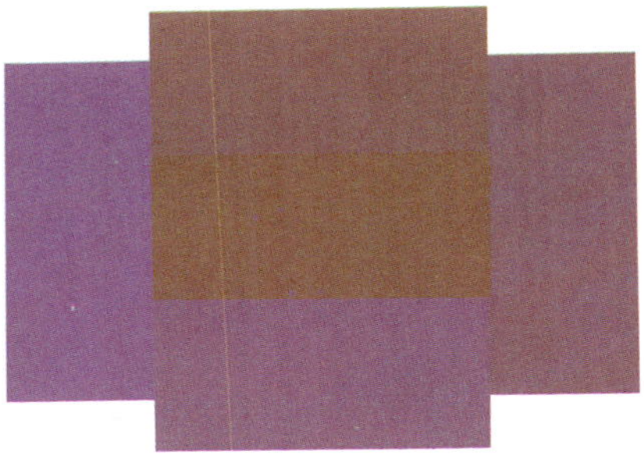

+100% PROCESS RED
+50% PROCESS BLUE

+50% PROCESS RED
+70% PROCESS BLUE

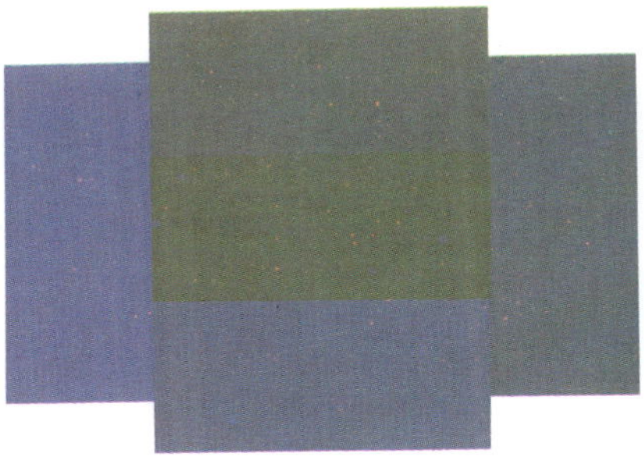

+70% PROCESS RED
+70% PROCESS BLUE

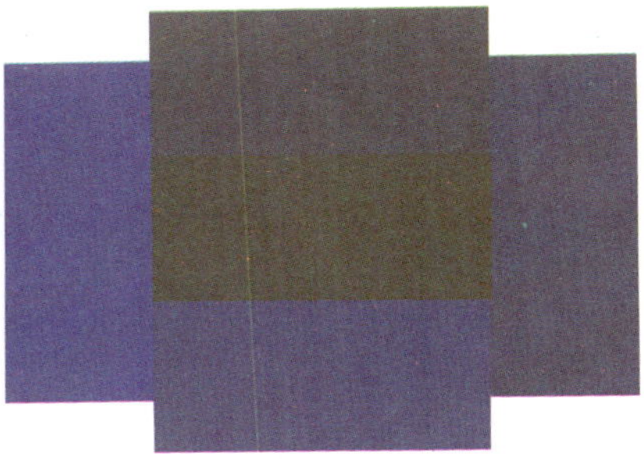

+100% PROCESS RED
+70% PROCESS BLUE

Key YELLOW tints
+10% BLACK

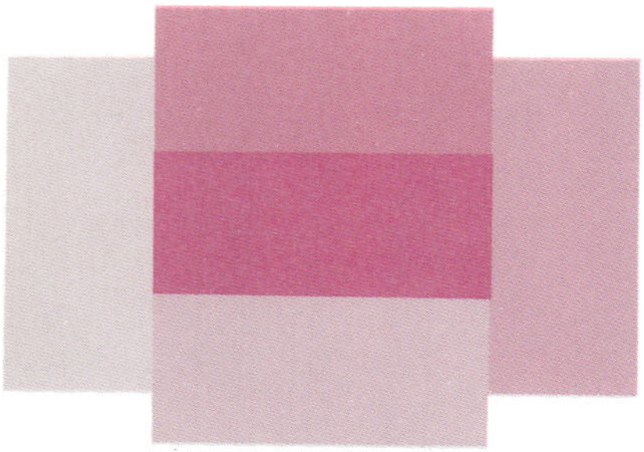

Key PROCESS RED tints
+10% BLACK

Key PROCESS BLUE tints
+10% BLACK

Key YELLOW tints
+30% PROCESS RED
+10% BLACK

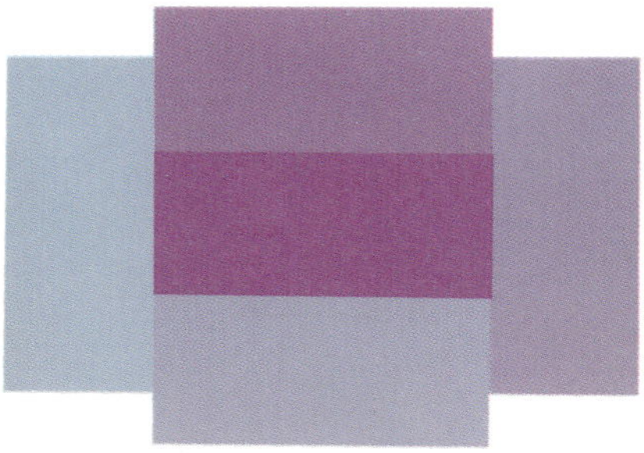

Key PROCESS RED tints
+30% PROCESS BLUE
+10% BLACK

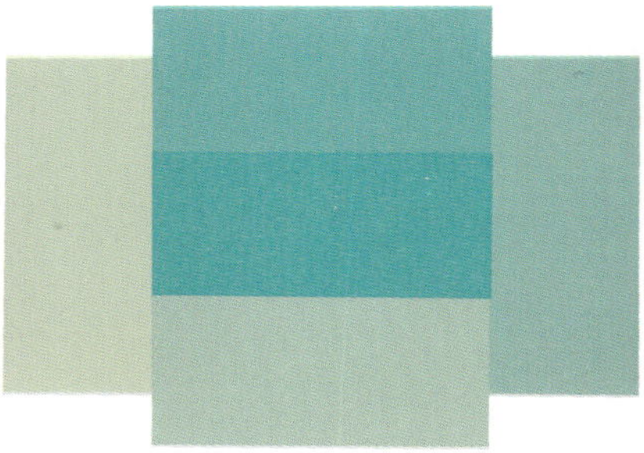

Key PROCESS BLUE tints
+30% YELLOW
+10% BLACK

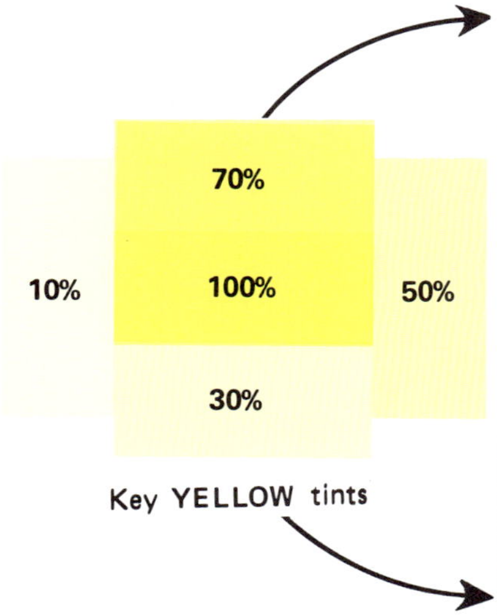

Key YELLOW tints

+10% PROCESS RED
+10% PROCESS BLUE
+10% BLACK

+30% PROCESS RED
+10% PROCESS BLUE
+10% BLACK

+10% PROCESS RED
+10% PROCESS BLUE
+30% BLACK

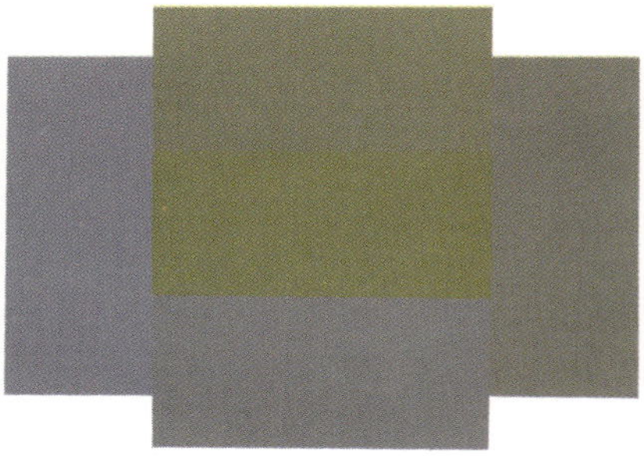

+30% PROCESS RED
+30% PROCESS BLUE
+30% BLACK

Key YELLOW tints
+30% BLACK

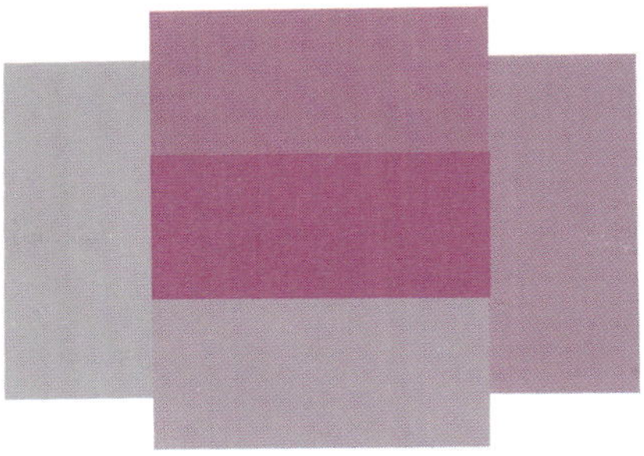

Key PROCESS RED tints
+30% BLACK

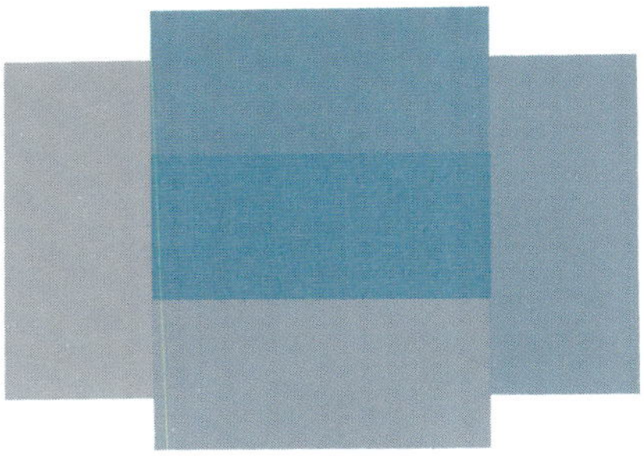

Key PROCESS BLUE tints
+30% BLACK

Key YELLOW tints
+50% PROCESS RED
+30% BLACK

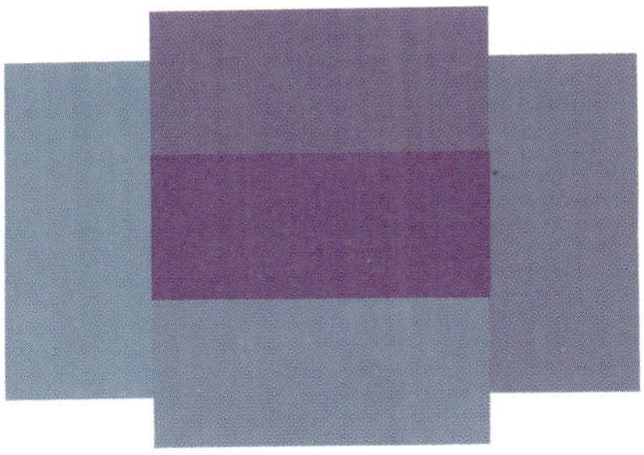

Key PROCESS RED tints
+50% PROCESS BLUE
+30% BLACK

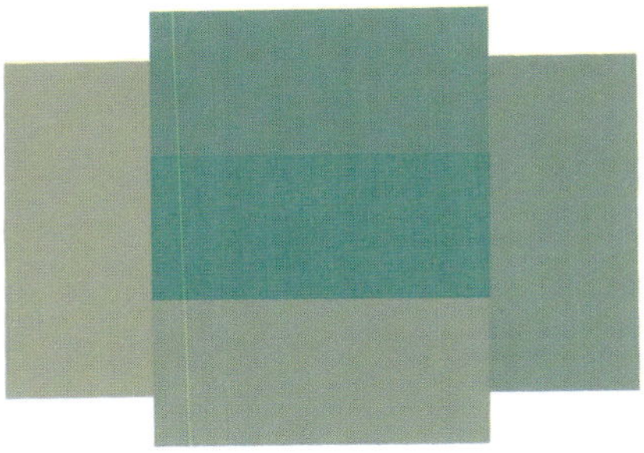

Key PROCESS BLUE tints
+50% YELLOW
+30% BLACK

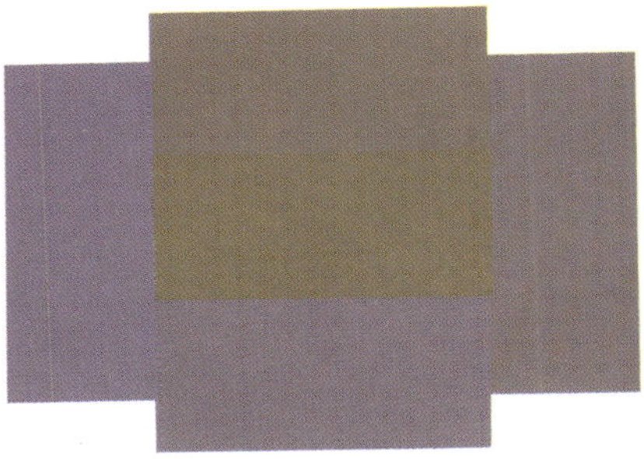

+50% PROCESS RED
+30% PROCESS BLUE
+30% BLACK

+70% PROCESS RED
+30% PROCESS BLUE
+30% BLACK

+100% PROCESS RED
+30% PROCESS BLUE
+30% BLACK

+50% PROCESS RED
+50% PROCESS BLUE
+50% BLACK

+70% PROCESS RED
+70% PROCESS BLUE
+50% BLACK

+100% PROCESS RED
+100% PROCESS BLUE
+50% BLACK

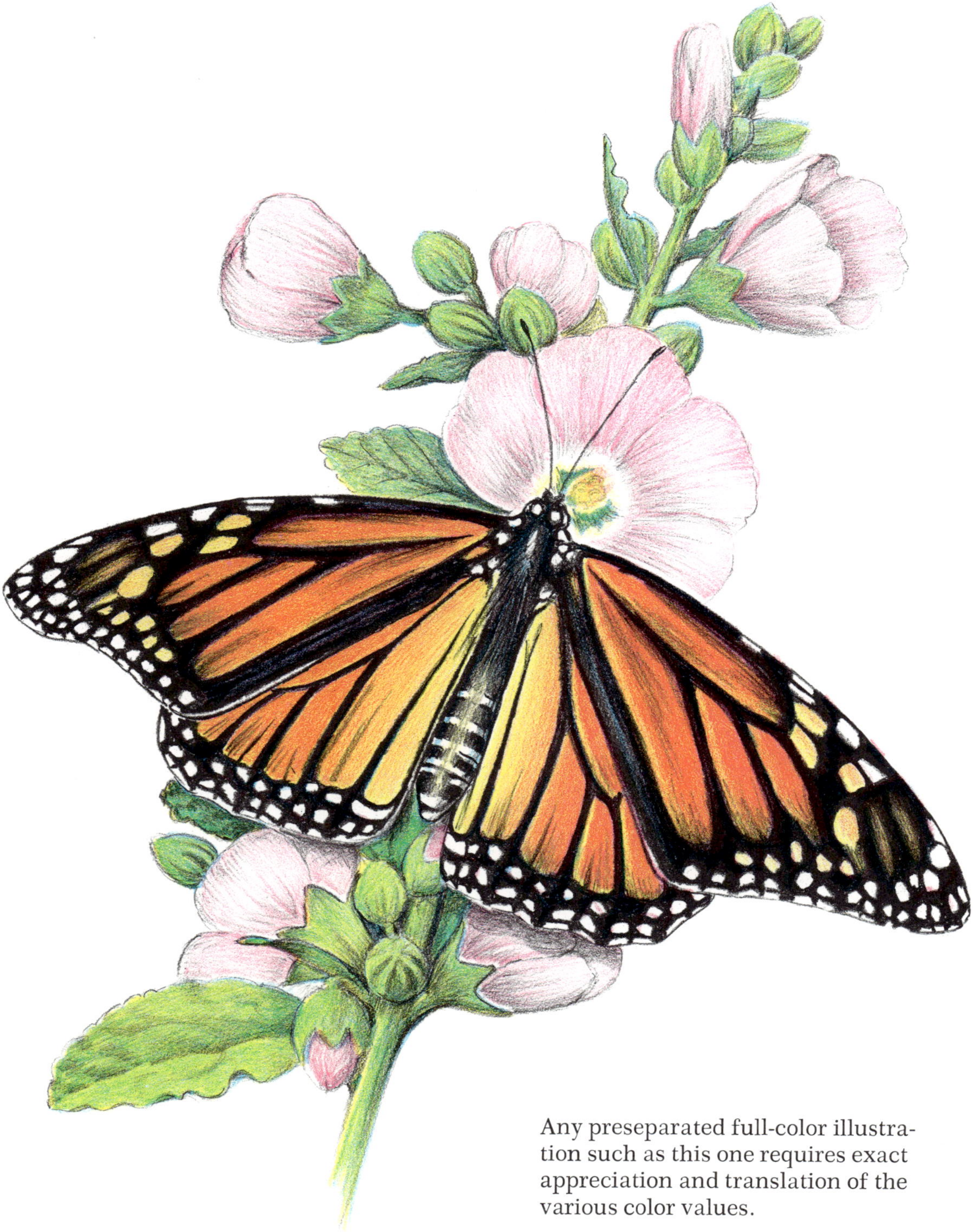

Any preseparated full-color illustration such as this one requires exact appreciation and translation of the various color values.

The Art of Separation

Preseparated four-color artwork is in reality a form of painting, even though it is rendered in black on four different sheets of film or acetate. It is a much more difficult and roundabout way of painting, which is not completed until it is reproduced in print because the printed image is its real goal.

The skills and abilities required to create such artwork go far beyond those needed for the average illustration. The skills are a matter of experience and perseverance; the abilities are native gifts, although they have to be cultivated and developed. One of those gifts is a sensitive eye for color, as indispensable to the painter as the pitch-sensitive ear is to the musician. The other is the ability to analyze the color in one's mind and to determine its components. Having an eye for color is not sufficient because the most important part of the preseparation technique is the analysis of each hue and the translation of the component colors into black values. Assuming that a brown hue is analyzed as containing 70% yellow, 40% red, and 20% blue, these values must be transferred in black to the four pieces of film or paper that make up what is known as the *camera copy*. Each of these pieces is labeled by the illustrator with the color in which it is to print, and receives a set of register marks. If the color analysis was correct, the four separations should suffice to recreate the original's coloring in the printed image—always provided that the printer's camera department is competent enough to reproduce accurately the artwork's values on film.

Obviously, the real difficulties involved in such a preseparation process are at the creative rather than at the printing end. Any experienced printing establishment should be able to produce the faithful negatives that are the basic requirement for a good reproduction. Upon the artist falls the full burden of providing carefully rendered separations whose reproduction in print will create an image combining perfect color detail with artistic composition.

Like any other ability, color perception develops and increases with constant practice, so that the analysis of color can eventually become almost an automatic reaction. Most persons looking at a picture of a brightly colored orange-and-black butterfly sitting on an aster will see it in terms of the attractiveness of the color combinations. They notice how the orange-brown tones of the butterfly are accentuated by the purplish hues of the flower and the green of the foliage. They would not normally, however, give any thought to the question of what color mixtures these hues represent, nor do they have any reason to think in those terms. But to the artist who must reproduce the components of all the colors on separations, the question of "How much of what do they contain?" quickly supersedes the initial appreciation of an attractive scene. The proportions of yellow and magenta that make up the orange in the butterfly's wings; the exact strengths of blue and magenta that yield the flower's purplish blue; the values of blue and yellow that add up to the green of leaves and stem; and last but not least, the degree of black that forms the shading and the butterfly's wing pattern—those are the questions that are important in the color analysis. All of them have to be resolved before work on the separations can begin, for corrections often seriously impair the quality of an illustration.

On the pages that follow, the process of rendering a set of balanced separations for the three primary colors and black is reproduced step by step. The butterfly on the flower was chosen as the model for this example because it offers a considerable range of different color combinations. This model is repeated opposite each individual separation so that color evaluation and analysis can be followed without difficulty. Apart from the black, which is the only one printing in its own color, all separations are reproduced twice: once in their own color, and once in black the way they are rendered on the transparent overlay for photography by the printer. This permits a comparison between the color values and their translation into black percentages.

Since many objects contain some admixture of black in the form of gray, and three-dimensional objects always have some shading that lends them depth and body, black makes a good key drawing for most preseparated four-color artwork. The black separation, as the only one that will reproduce in the same color in which it is rendered, provides a kind of skeleton shadow image to which the three primaries may be added one by one in appropriate strengths.

By choosing to render the black drawing first, the illustrator is starting out with the image that may come last when the artwork is reproduced on press, where the black is often imprinted over the combined yellow-red-blue image. It is well to keep this in mind in order to avoid making the black so strong that it will overwhelm and muddy the various chromatic hues of the illustration.

Once the black key drawing is finished, the other separations may be done in any sequence desired by the illustrator. Usually it is best to have at one's disposal a color guide in the form of a sketch or a photograph made expressly for that purpose. The task of correctly gauging the various hues is formidable enough even while looking at the original without adding to it by relying solely on memory. In some instances it may of course be possible to have the original object as a model throughout, and so be rid of the need for sketches or photographic color aids.

The selection of yellow as the second separation in this instance was based on the fact that considerable amounts of this color are contained in all parts of the picture, with the exception of the petals. In determining the various degrees of yellow in the mixed hues, the component colors have to be subtracted. By mentally removing the magenta from the orange areas of the butterfly's wings, and the blue from the green of the leaves, it quickly becomes clear that the yellow content of both is high; the flower centers also require a strong yellow tint. In rendering the yellow as black percentages on the separation, one has to remember that there is an unconscious inclination to keep the tones lighter than they should be because yellow is such a light color, and the mind subconsciously

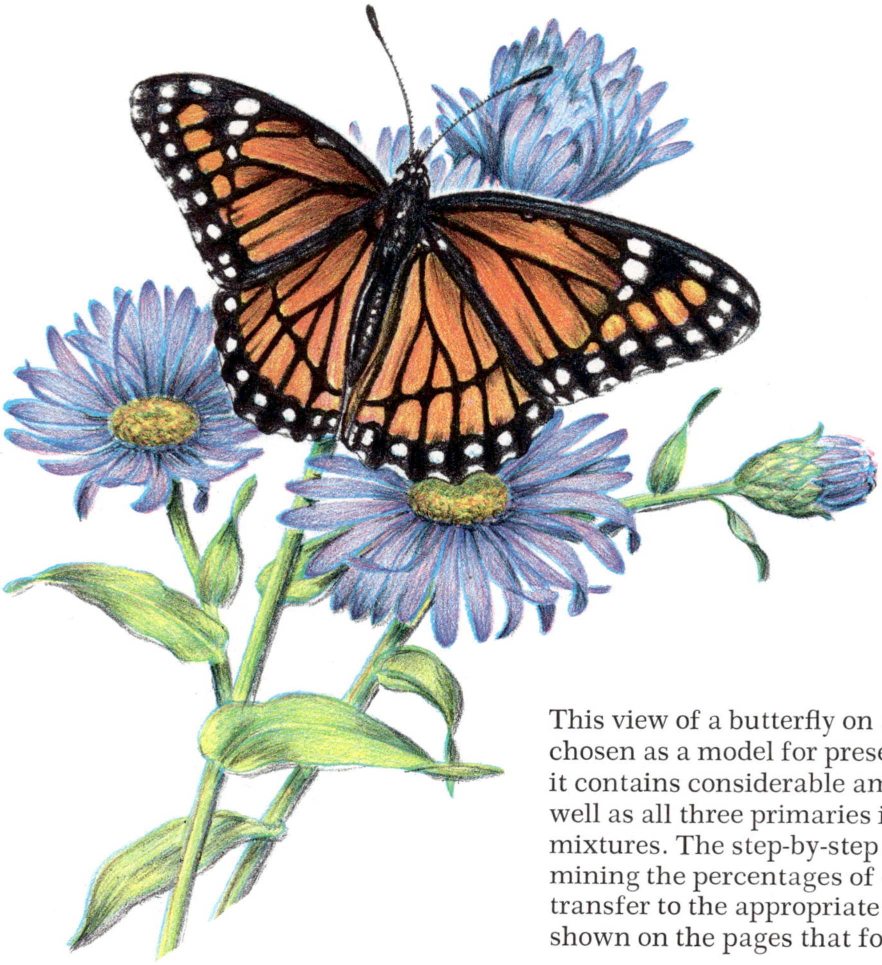

This view of a butterfly on an aster was chosen as a model for preseparation because it contains considerable amounts of black as well as all three primaries in different mixtures. The step-by-step process of determining the percentages of each color for transfer to the appropriate separation is shown on the pages that follow.

rebels against using so much black to reproduce so sunny a hue. A quick look at the color charts, however, will be reassuring by showing how much yellow is needed to produce a deep, strong orange.

As a final touch on the yellow separation, a light tint is added to the black wing and body areas of the butterfly. Reinforced by similar tints of the other colors, this will make the black look alive and give it a properly velvety appearance.

Magenta becomes the third separation. Both the butterfly and the flowers clearly contain a good deal of red. However, it should be remembered that magenta, a mixture of red *and* violet-blue, will cancel out all the green in the yellow if the two are combined in full strength. In order to create a real orange, which contains green, the percentage of magenta must be less than that of the yellow.

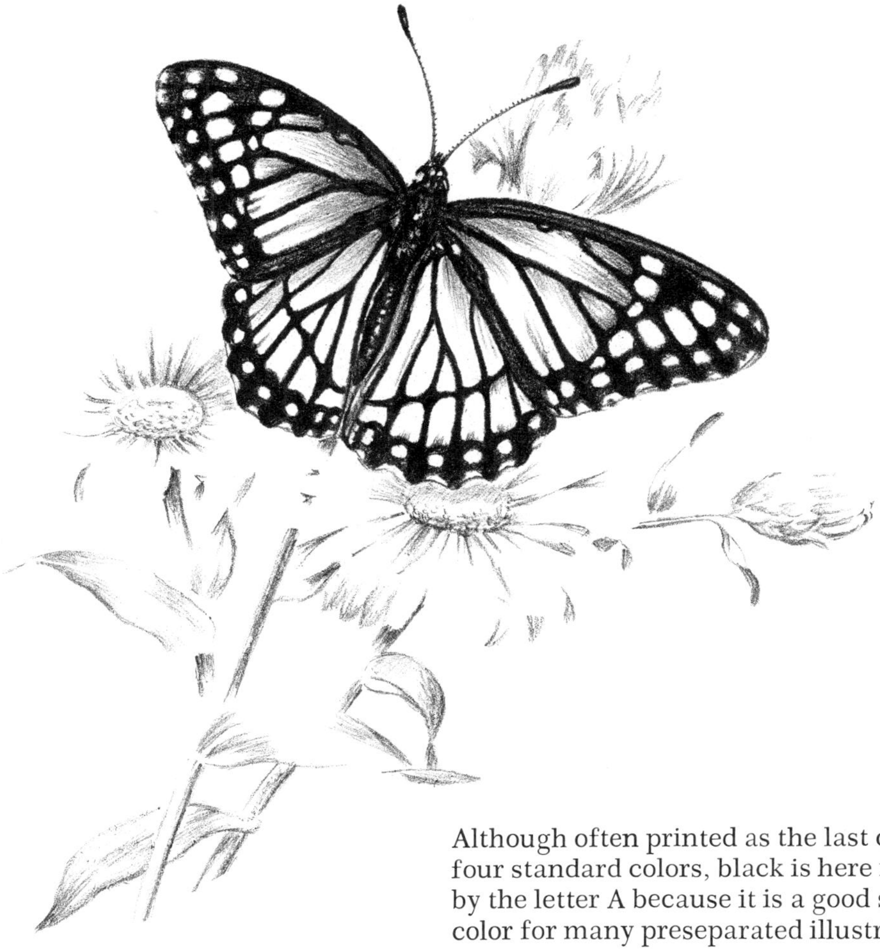

A

Although often printed as the last of the four standard colors, black is here marked by the letter A because it is a good starting color for many preseparated illustrations.

It is a different matter with the flower itself. The violet-purplish hues of the aster contain practically no green at all, so the magenta must be strong enough to cancel out all but a slight trace of the green present in the blue that will be added on the last separation. The flowers therefore have to receive the exact tint of magenta capable of achieving that result without making the hue too reddish. Again, a look at the color charts will be helpful.

Finally, a very light tint of magenta provides the flower center with the proper shade of deep yellow, and even lighter touches added to the green of the leaves will keep the green from having an unnatural "grassy" appearance. It is remarkable how many of the green shades in nature contain varying amounts of red, even though this is often hardly noticeable.

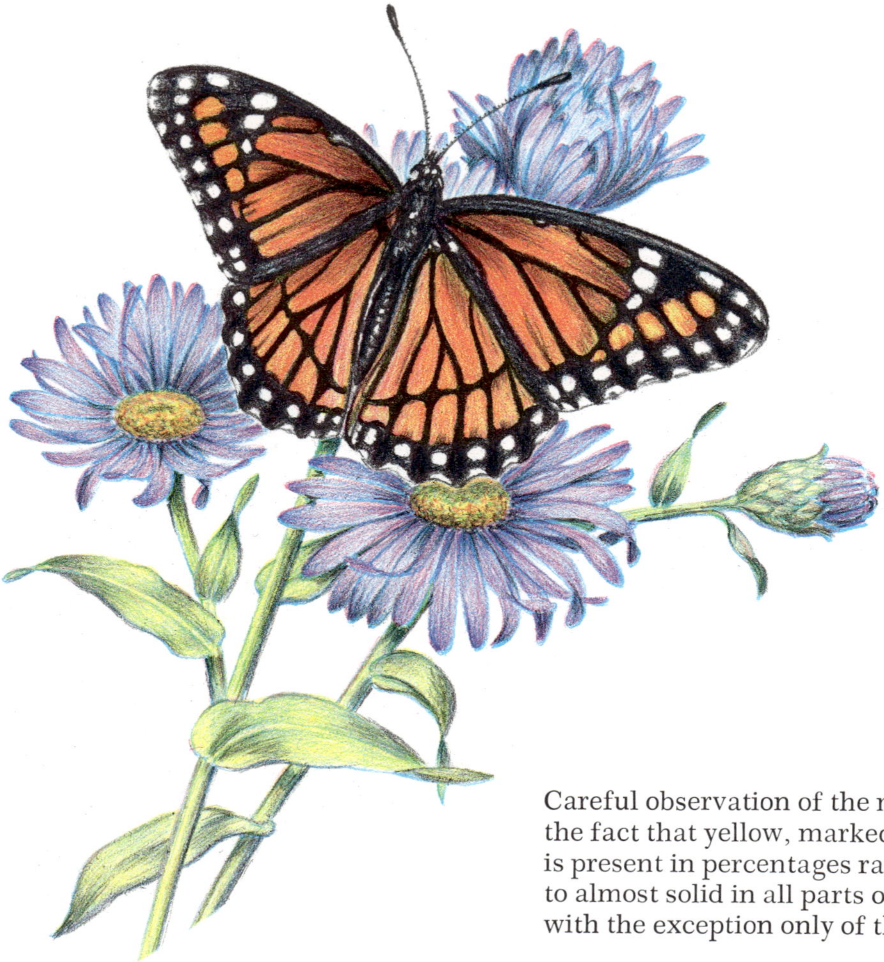

Careful observation of the model confirms the fact that yellow, marked by the letter B, is present in percentages ranging from light to almost solid in all parts of the original with the exception only of the petals.

A light tint of magenta on the black wing areas finishes this separation; these color tints underprinting the black are very important for correcting the deadening effect of the neutral black ink in the reproduction.

Blue is the last of this set of separations; like the other colors, blue is contained in many parts of the illustration. Only the orange wing areas get no blue at all. The color is strongest in portions of the flower heads; even there, however, it is not used in full strength. This applies to the other colors also, although yellow and black come close to full strength in some parts of the butterfly. The rest of the illustration is a combination of delicately graded tints.

Because the asters are more violet than purple, the blue tint must be deeper than that of the magenta for the same area. Not

B

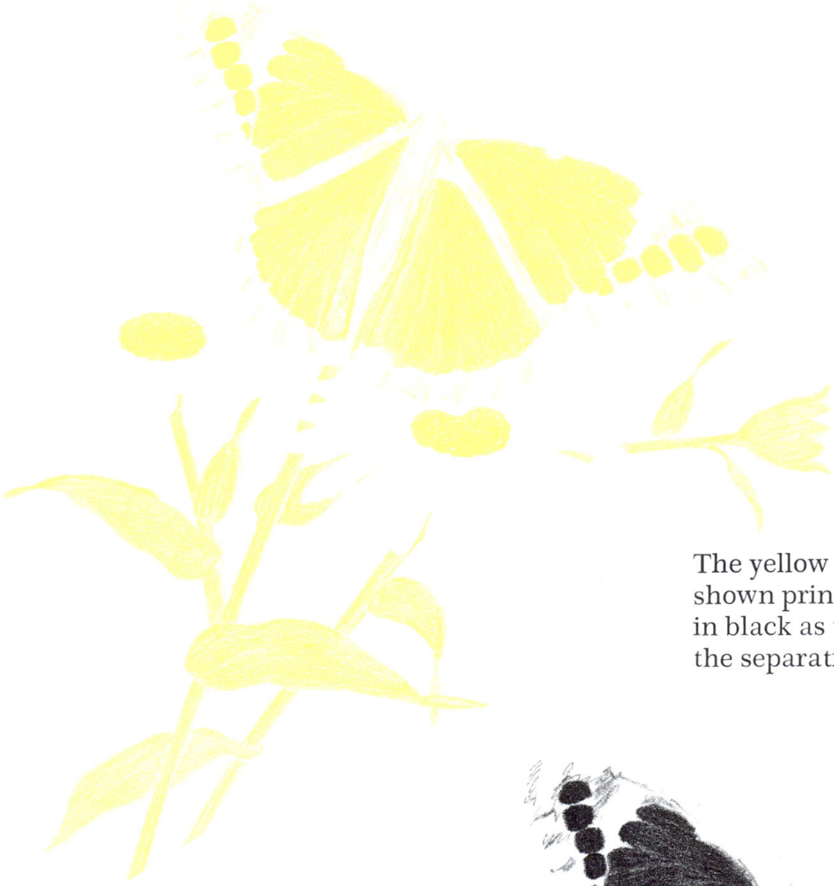

The yellow percentages shown printed in yellow, and in black as they appear on the separation.

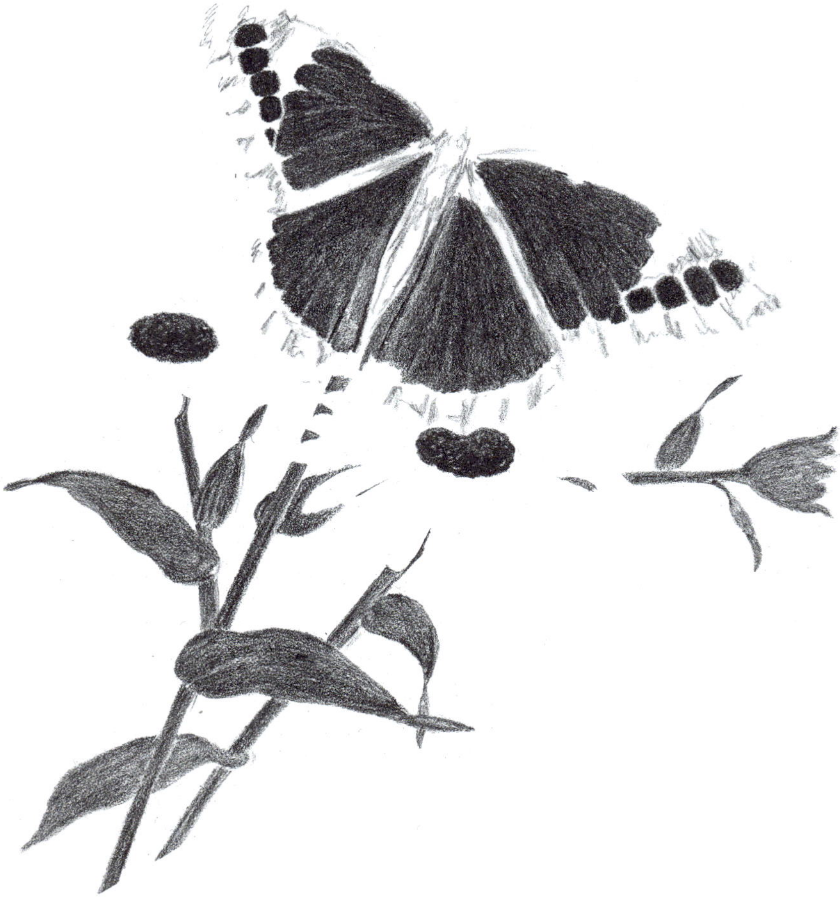

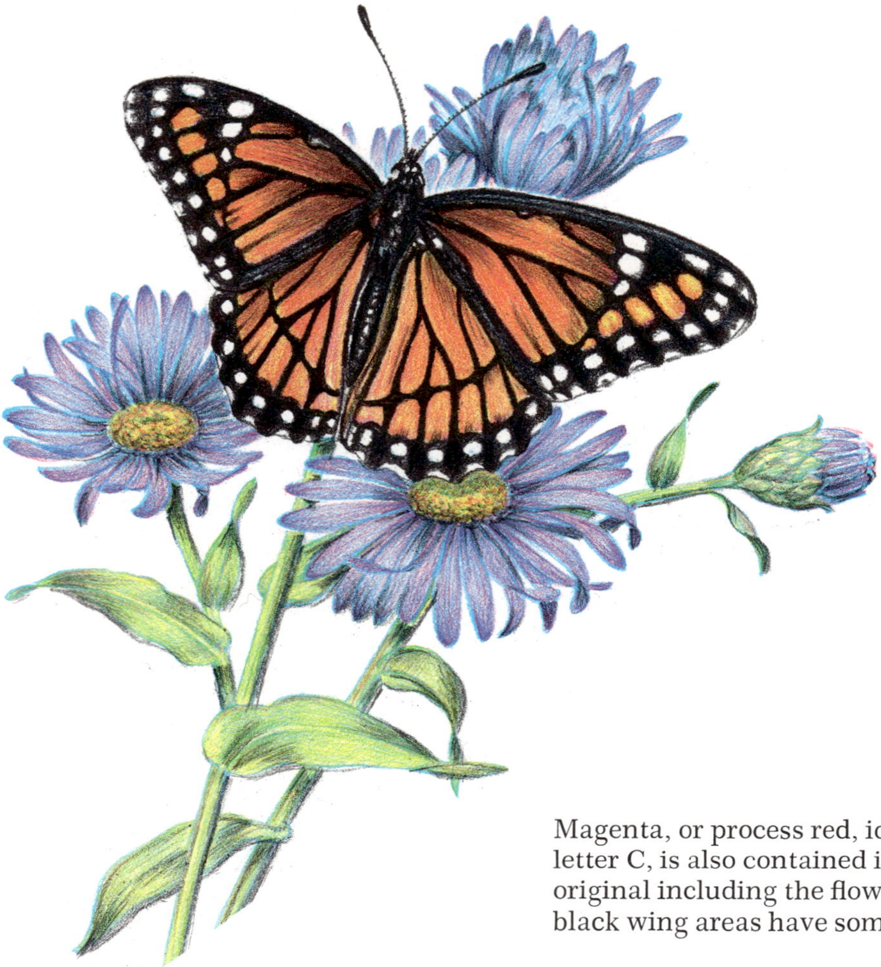

Magenta, or process red, identified by the letter C, is also contained in much of the original including the flower heads; even the black wing areas have some red.

too much deeper, though, or the green content of the blue will prove overpowering. For the leaves also, the blue will have to be used with discretion; a look at the blue-yellow combinations of the charts confirms the fact that process blue is an extremely strong color. The percentage needed to produce the right green in this instance is only about half of what was required for the same area on the yellow overlay.

Once again, a light tint is added to the black wing parts on this separation, and that final touch completes this set of four black line illustrations that are supposed to re-create the original model in a printed image that has a delicately shaded "halftone" effect.

There are few things more fascinating to watch in printing than the piecing together, color by color, of a four-color illustration.

C

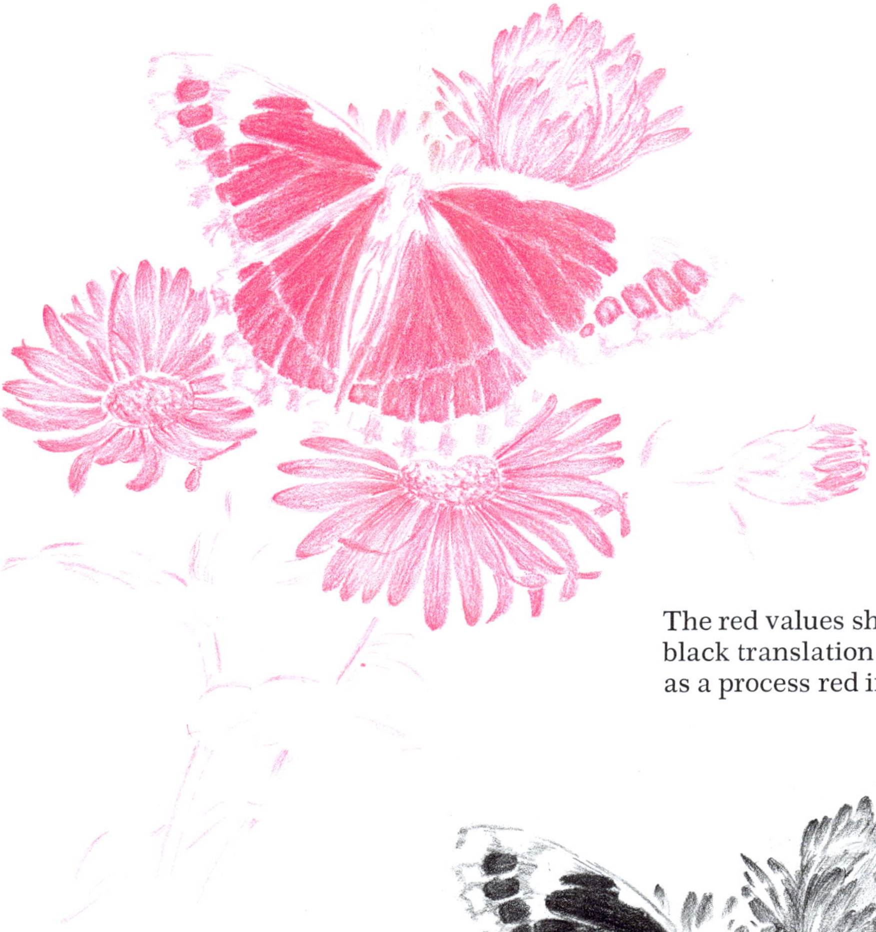

The red values shown in their black translation below and as a process red imprint.

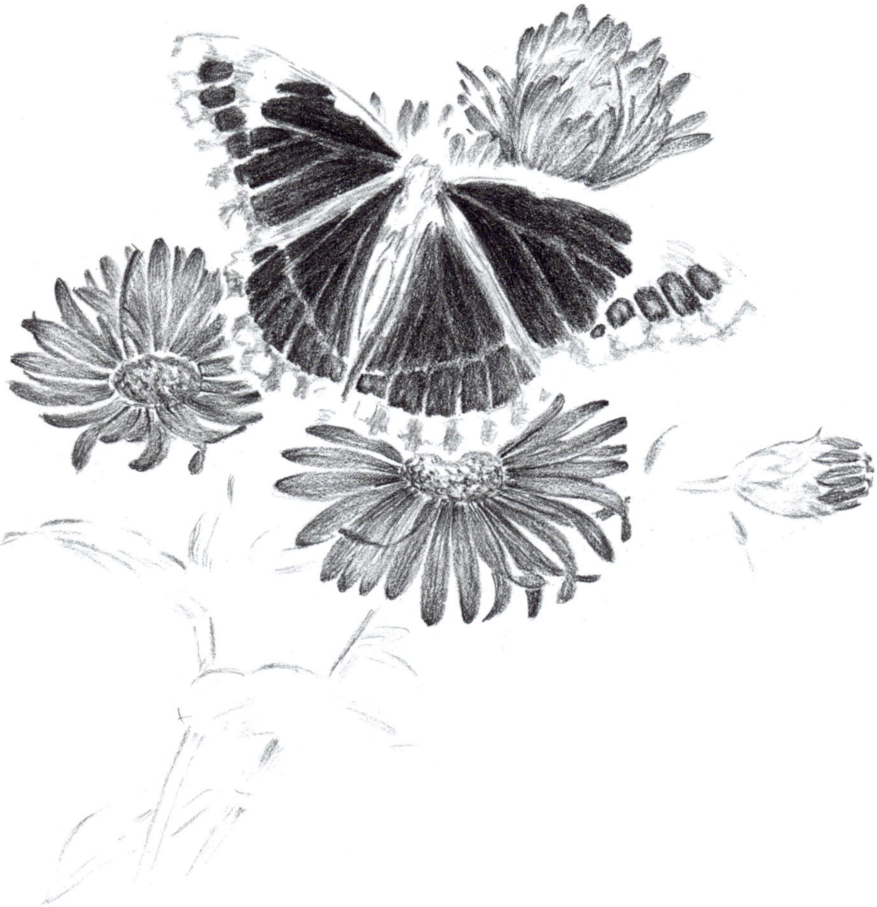

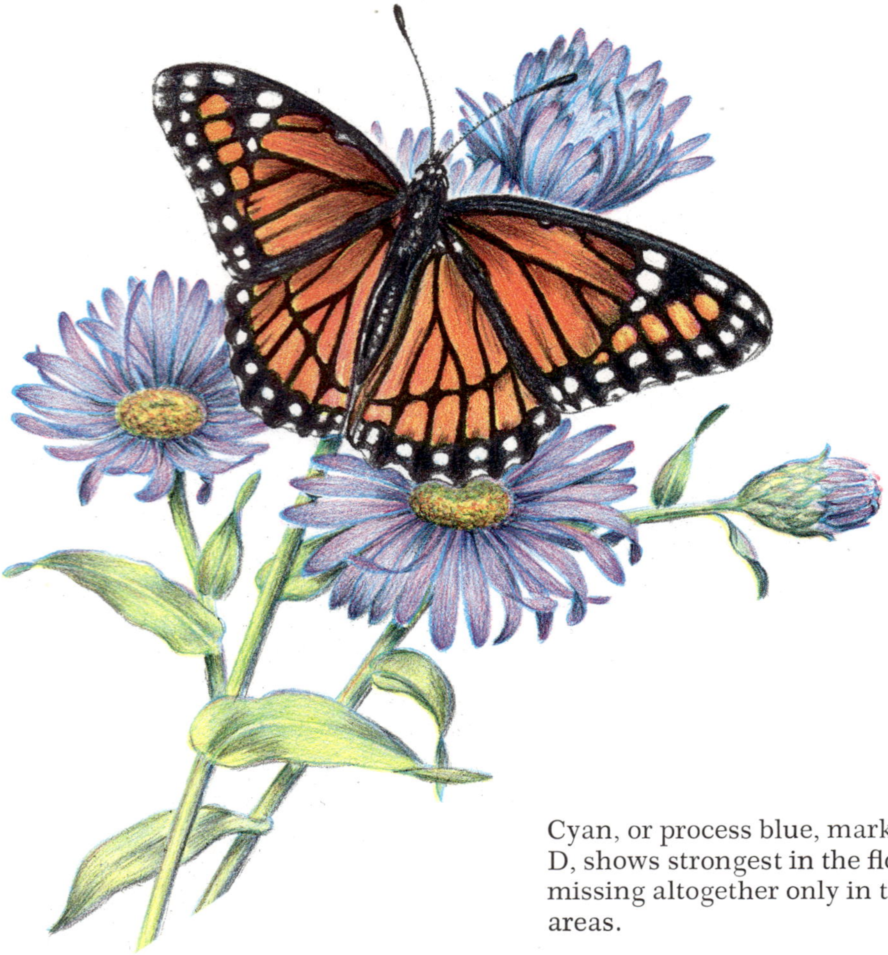

Cyan, or process blue, marked by the letter D, shows strongest in the flowers and is missing altogether only in the orange wing areas.

The series of four single-color images plus the combination of first two, then three, and finally all four colors is known as a set of progressives. First, the magenta impression is printed over the yellow. Then the blue impression is added, and lastly, the black "skeleton" is superimposed over the combined three primary images. The reproduction is now complete and should resemble the original in all particulars.

This step-by-step demonstration is reproduced on these pages mainly because it provides visible proof of several points made in earlier chapters. Both in the two- and three-color image, for instance, the orange wing areas do not look nearly as brilliant as those in the four-color image. The reason is not that the orange has changed but that the black wing veins and borders enclose it

D

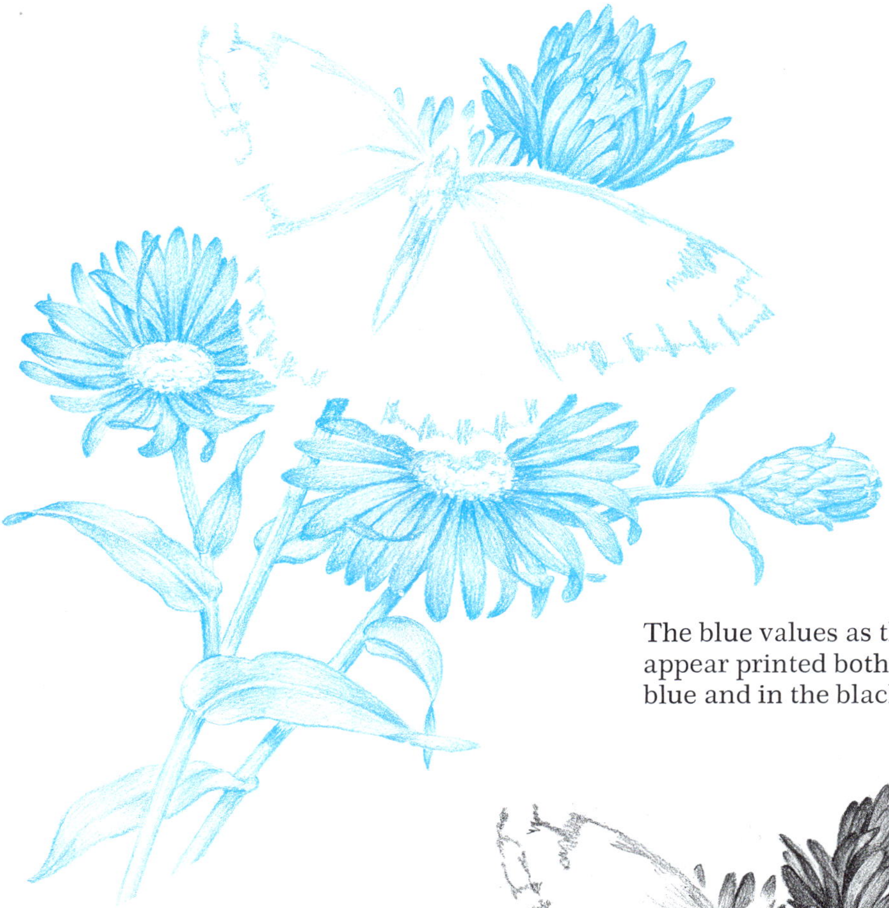

The blue values as they appear printed both in process blue and in the black translation.

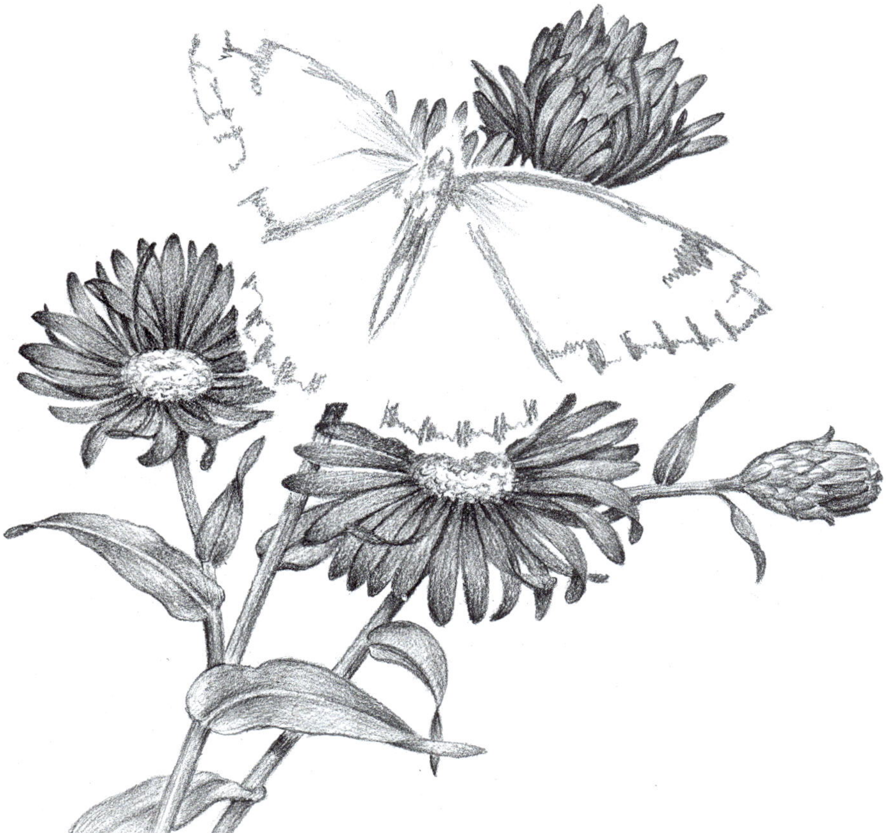

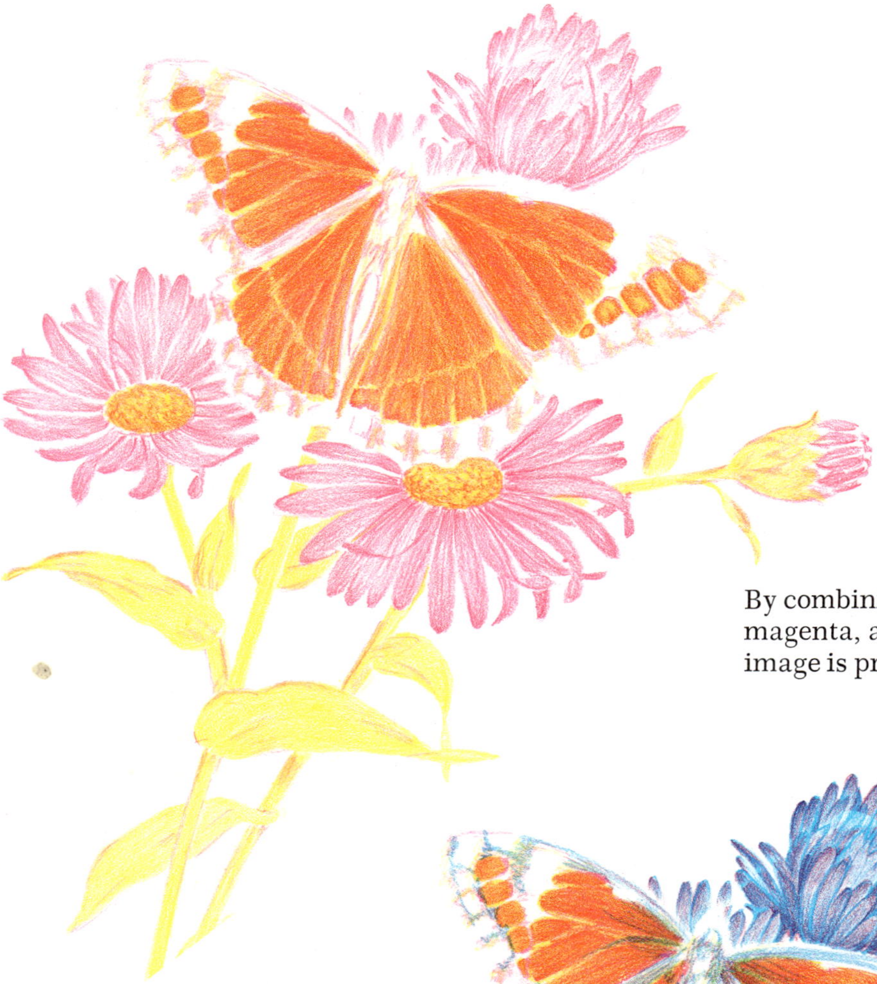

B+C

By combining yellow and magenta, a two-color printed image is produced.

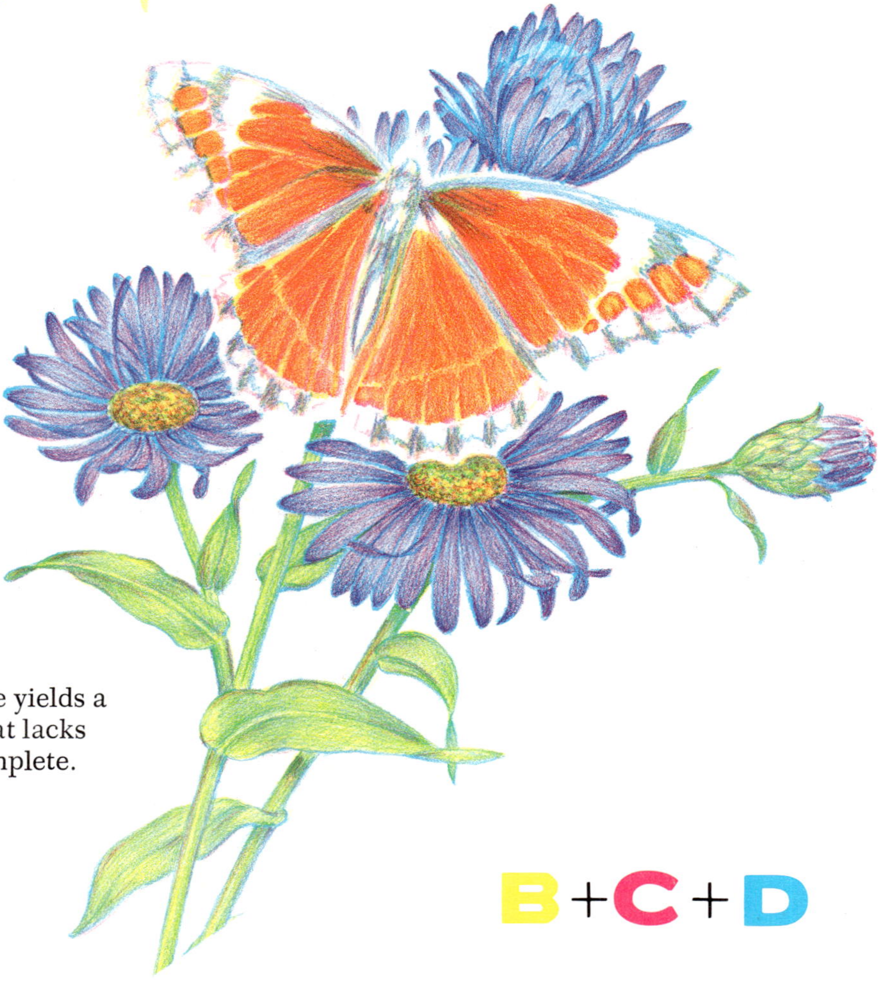

The addition of blue yields a three-color print that lacks only black to be complete.

B+C+D

on the final print, thereby creating a sort of "stained-glass" effect that accentuates the brightness of the color.

On the other hand, the flowers seem almost overpowering on the three-color image, but are immediately reduced to their proper strength by addition of the black, and so offer proof of how strongly colors are affected by their neighbors. A paler violet would not have been able to hold its own against the strength of the black. By the same token, if the butterfly had been a pale and weakly patterned species, a somewhat more subdued flower color might have been a better choice. The freedom to use such refinements and adaptations is one of the factors that make the technique of four-color presepa-ration so sensitive a medium. Employed with skill and imagination, it is capable of creating sophisticated artwork precisely geared to producing superior full-color effects in the printed end result.

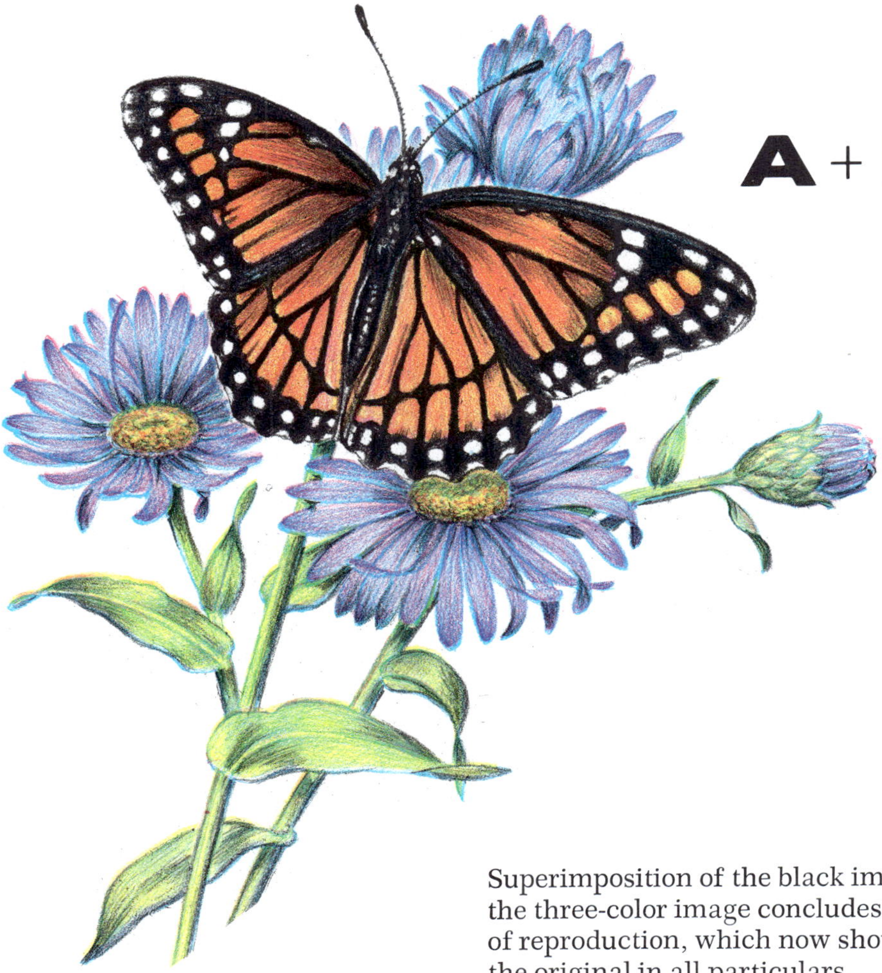

A + B + C + D

Superimposition of the black image A upon the three-color image concludes the process of reproduction, which now should resemble the original in all particulars.

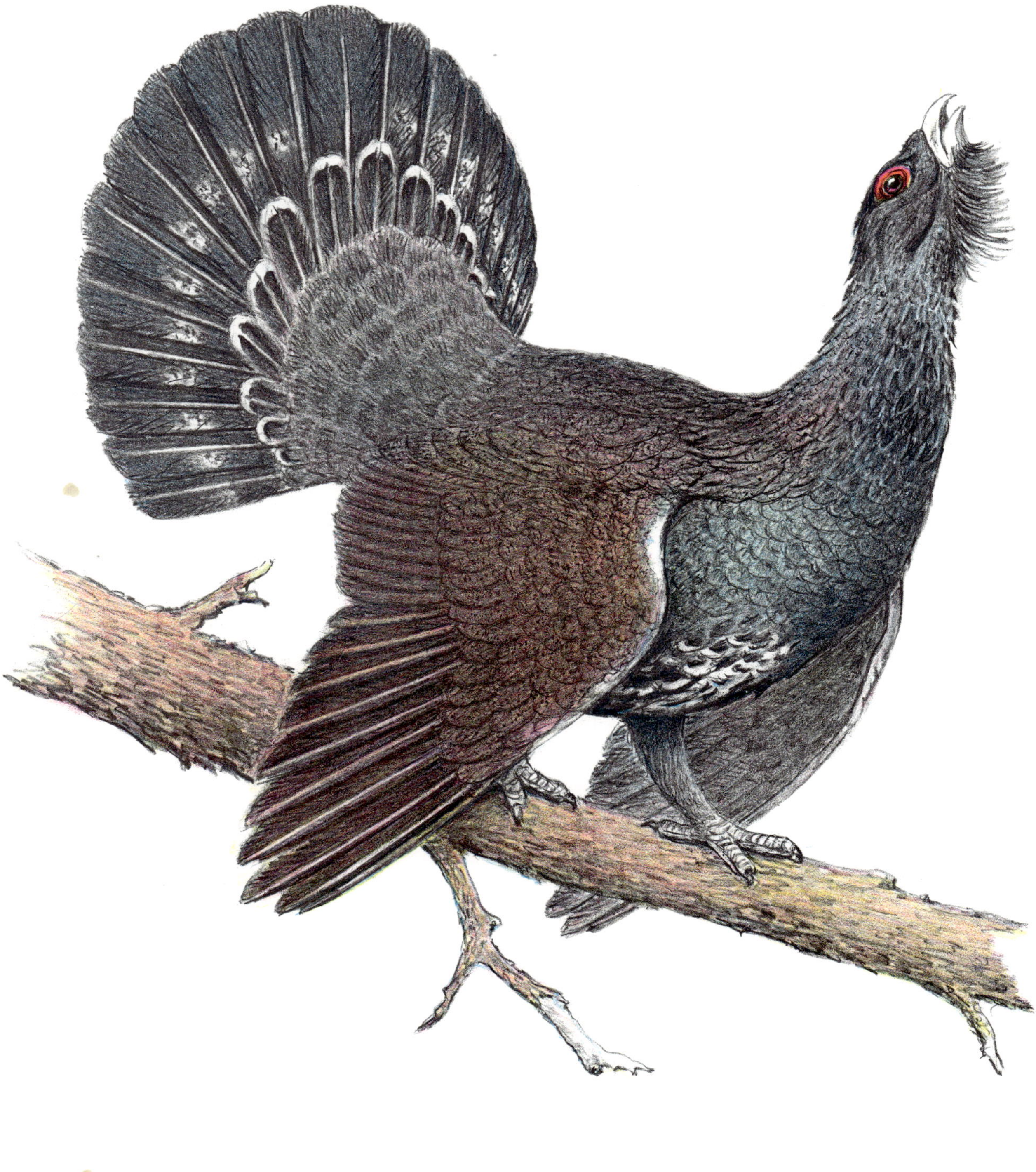

The subtle coloring of this bird's plumage, which required mixtures of all four printing colors, demanded great precision in color evaluation.

Adventures and Challenges

My first attempt at writing and illustrating a book was for a story grouped around more than two dozen full-page watercolor illustrations featuring plants and animals. They displayed all the colors of the spectrum, and hues ranging from the most delicate to the brilliant. Still in my teens when I began working on that project, I had no knowledge of reproduction methods and no criterion of artwork other than a purely aesthetic one.

When I finally submitted the manuscript and the illustrations to a major New York publisher, it turned out to my intense disappointment that the illustrations, which I had considered the book's greatest asset, were the stumbling block to its acceptance for publication. The editor of the juvenile department explained that the cost of this full-color art would be prohibitive for the first book of an unknown young author. Although she initially suggested that I be satisfied with a full-color jacket and redo the illustrations in black only, she finally had to admit that color was essential to my text.

After that first great disappointment I was even more determined to succeed and unwilling to limit myself to subjects that would not require colored illustrations. I therefore made up my mind to learn as much as possible about the mechanics of the process by which artwork is turned into printed reproductions, hoping that such information would eventually permit me to produce colored illustrations that would be both attractive and acceptable to an editor. Since I also needed to make a living, I decided to accept a position in the art department of a publisher specializing in children's books.

It was part of my job there to handle a great many illustrations rendered in a variety of techniques. I learned about the different types of line drawings, came to appreciate the difference between line and halftone, and found out that "full color" in today's printing idiom almost always means four colors. Eventually, it became clear to me why full-color art reproduced by photomechanical "process" separation is so expensive. The opportunity to compare, time and again, the printed image of such art with the original opened my eyes. I soon realized that some loss of detail and definition was unavoidable, and was surprised at the often extensive color adjustments and corrections that were necessary to bring the reproduction closer in appearance to the original. Occasionally the extent of such corrections became the subject of lively arguments between publisher and printer.

By observing and absorbing all these nuggets of information, I gradually built up a fair amount of useful information not only about the process of reproduction but also about the advantages—and limitations—of the various art techniques commonly used for illustrations.

The day arrived when the editor asked me to try my hand at

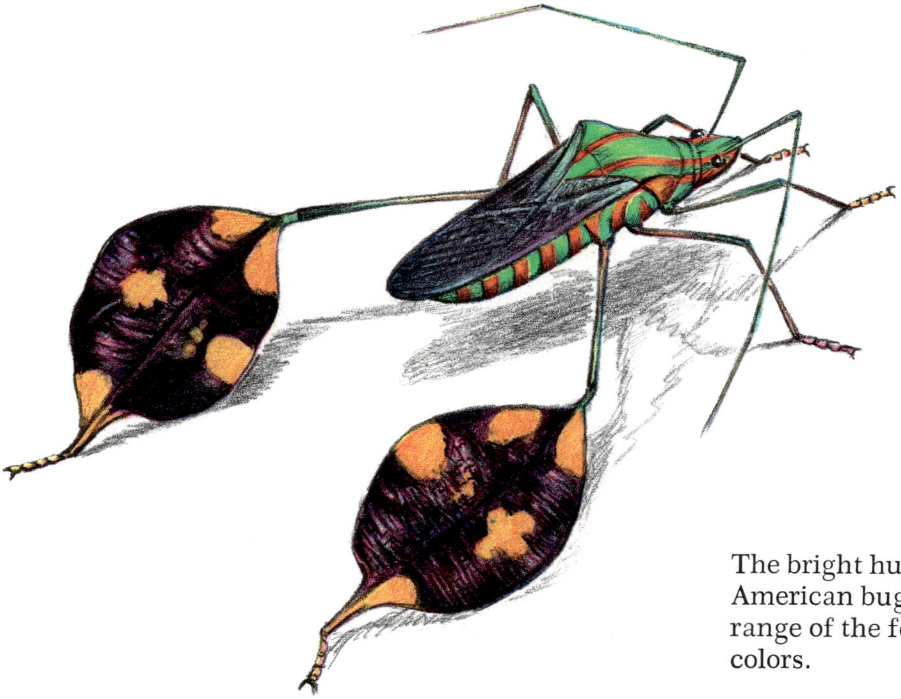

The bright hues of this South American bug span the full range of the four printing colors.

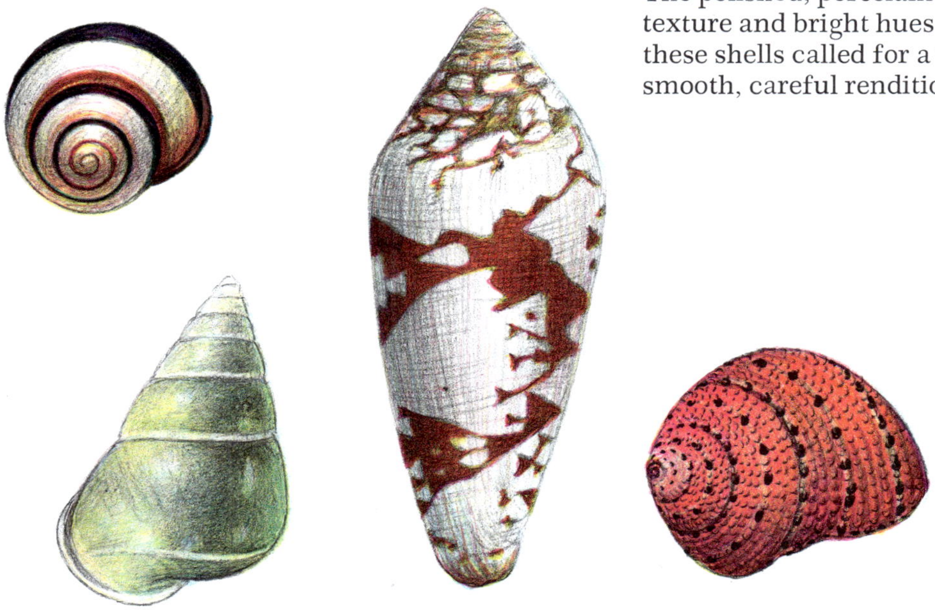

The polished, porcelain-like texture and bright hues of these shells called for a very smooth, careful rendition.

preparing single-color separations, or overlays, for some black line drawings; formerly, this work had been handled by a free-lance artist. Although I had never before done anything like it, the overlays did not seem difficult. The areas designed to print in color had to be either inked in with black for solids on acetate sheets or covered with preprinted adhesive screen tints for the lighter shades.

Working on the separations, I found it surprisingly easy to imagine exactly how black or a shade of gray would look in color, and soon began to amuse myself by devising more imaginative and more finely graded combinations. When the printed sheets of the first book featuring my overlays arrived at the office from the printer, there was a ripple of excitement in the editorial department. Afterward the president of the firm congratulated me in person, and as a more concrete sign of his satisfaction surprised me with a raise in salary.

Although delighted by that highly welcome raise, I was even more gratified by the success of my first experiment, for those single-color separations had started a new train of thought in my mind. Granted, the technique of adding color with solids and flat screen tints was rather crude—but what if the same concept could be adapted to a very subtle and delicate technique? I was already thinking in terms of more than one color, for there seemed to be no reason why it should not eventually be possible to find a way of rendering separations in several colors for full-color illustrations. The ease with which I was able to translate color into black values and vice versa made me feel confident about such a possibility. All I needed was some more experience and a few crucial experiments to find a viable preseparation technique.

The opportunity to test one possibility came when I volunteered to do the illustrations for a book about birds with which the publisher was having some trouble because it called for an artist with some ornithological knowledge. I had more than enough of that, and was very glad of the chance to experiment in black and white with a technique that I someday hoped to apply to full-color illustrations. Working with pencil on a fine-grained paper, I produced drawings that looked promising. But when the printed sheets arrived, the result, although very satisfactory to the publisher, was disappointing to me. The screen of the halftone had robbed my drawings of fine detail and sharpness of line, and some of the shaded areas looked fuzzy and flat.

Now began the search for a material that would make screening unnecessary yet produce a halftone effect. Finally I came across a translucent film used mainly by architects and engineers. It had a finely stippled surface, and responded well to a suitably soft pencil. Drawings executed on that material produced images with delicately shaded areas that looked like halftone but in fact were line, and could be reproduced as line by direct contact negatives without any need for halftone screens. Here, at long last, seemed to be a medium well suited for what I had in mind; my goal, although still a long way off, seemed attainable.

A big step in the right direction came with the publication of my first book tailored to publishing needs as I had learned to understand them. Featuring social insects, and illustrated with preseparated two-color art rendered on the stippled film, the book was promptly accepted by a New York publishing firm. It reproduced extremely well and was singled out for praise by reviewers. I now

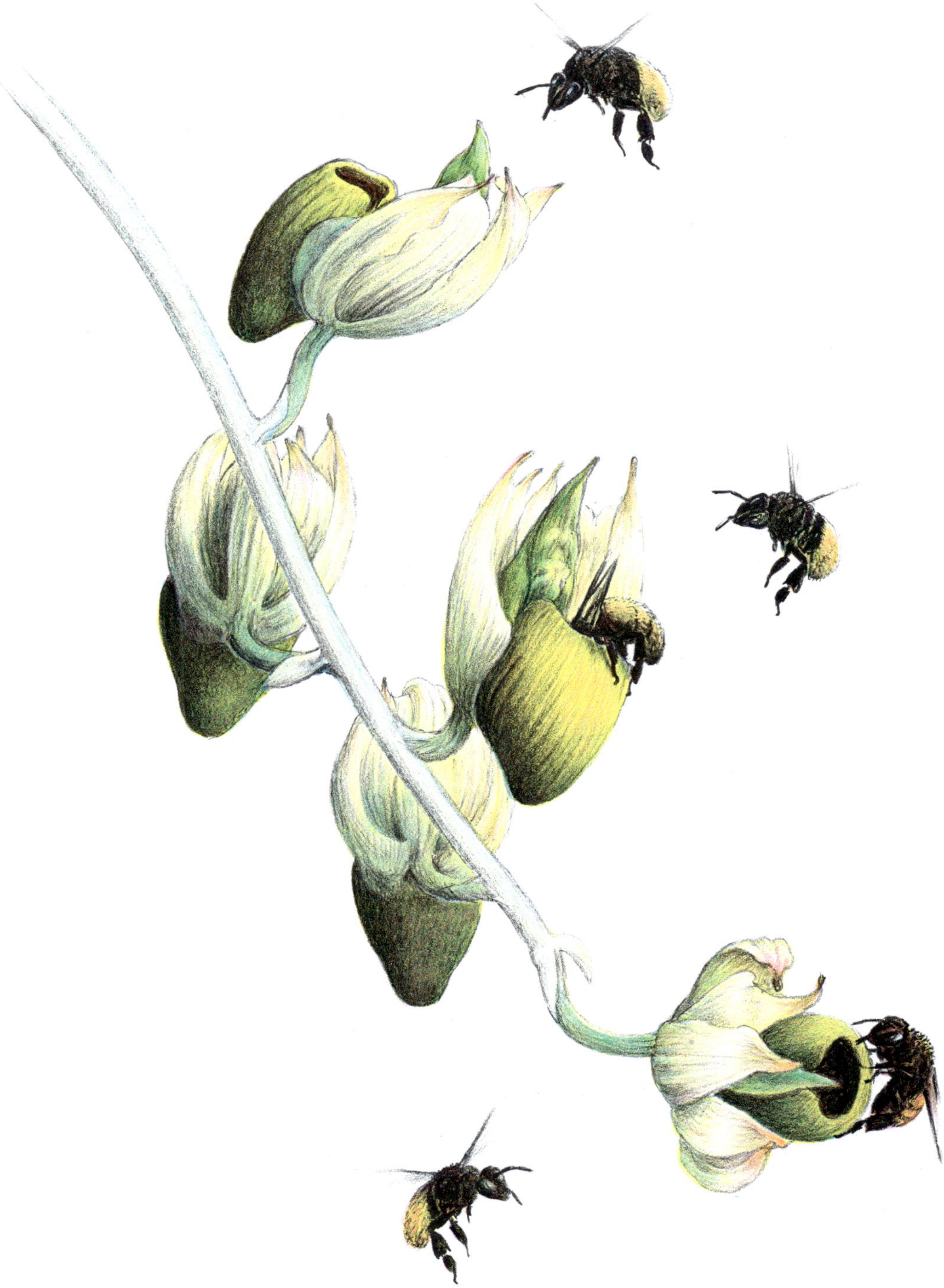

Subtle but distinctive, the delicate tints of this catasetum orchid presented a challenge to preseparated line rendition.

was ready for the most important test; by sheer luck, the printer who had handled the insects book had become so interested in my technique that he offered to make, free of charge, a sample proof of a four-color preseparated illustration I had prepared for that purpose. The result was all I had dared to hope for. The long search was over; all the colors in that reproduction had the right values. Both the monarch butterfly and the flower in the illustration looked alive and glowing with color. It seemed clear that this technique could be perfected and adapted to a wide variety of subjects.

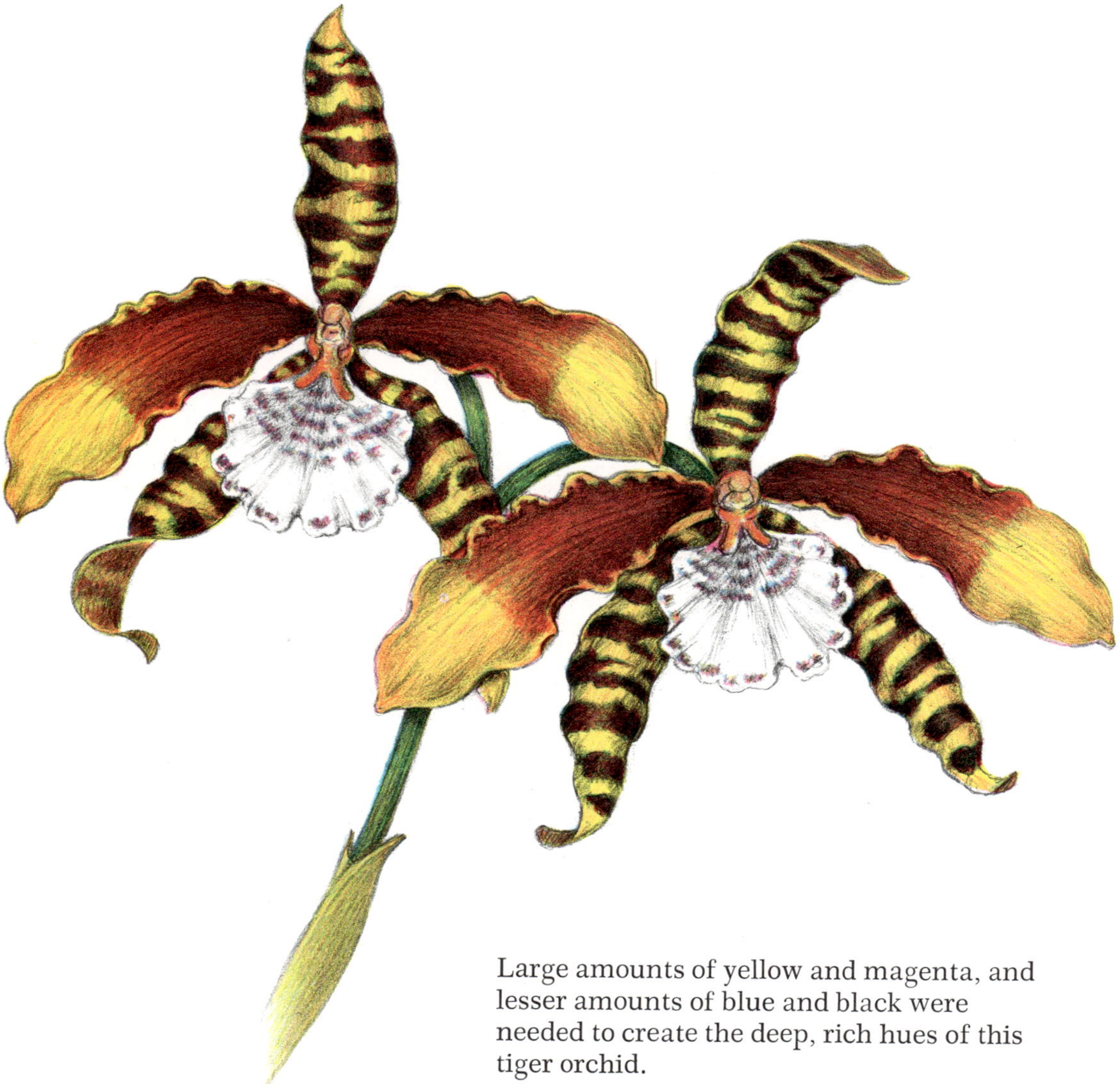

Large amounts of yellow and magenta, and lesser amounts of blue and black were needed to create the deep, rich hues of this tiger orchid.

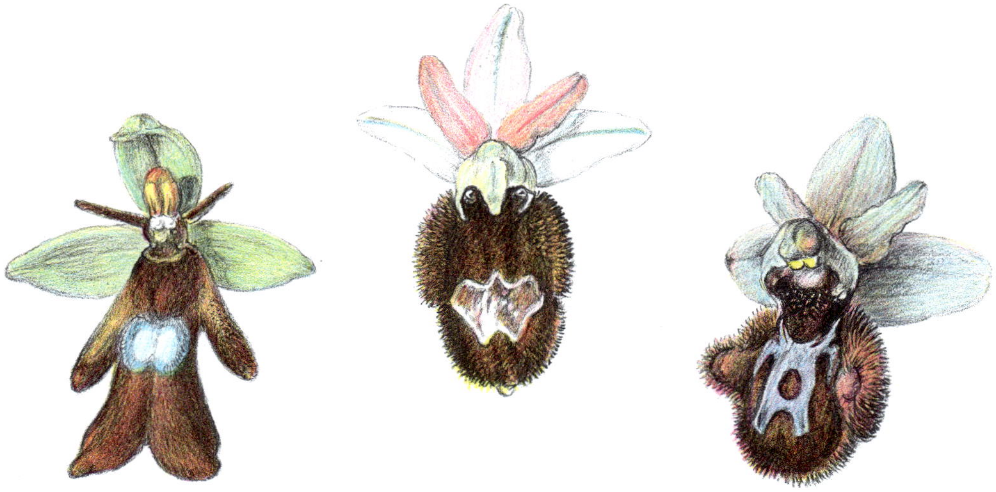

Tints of all four printing colors ranging from the lightest to almost solid were required for these ophrys orchid blossoms.

The task of convincing others of that fact was quite another matter. Although my first book illustrated with four-color separations was accepted not long after by another publishing firm, the people in the production and editorial departments were so jittery about my strange technique that they insisted I do a conventional watercolor at least for the jacket, even though that meant an expensive process separation. Matters were not helped much by their printer's assertion that my separations would never produce the results I was claiming for them. A sample proof of one illustration finally convinced all the Doubting Thomases, including the printer; the book reproduced very well indeed and even collected an award.

From that time on, I had no difficulty getting my preseparations accepted by publishers; on the contrary, they seemed to appreciate greatly the advantages of such illustrations. For me, that meant a good deal of freedom to experiment, in each additional book, with new ways of improving and broadening the scope of my technique.

Light tints of magenta and blue added to strong but sculptured black and yellow percentages created the shiny appearance of this brass powderhorn from the days of the American Revolution.

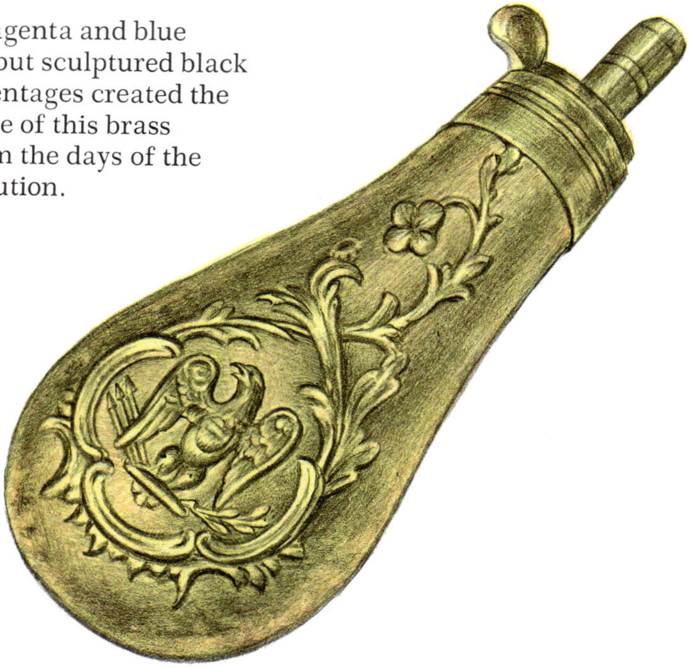

During the first few years I had been using it almost exclusively for the rendition of plants and animals because all my books dealt with such subjects. Then one day a friend of mine asked me

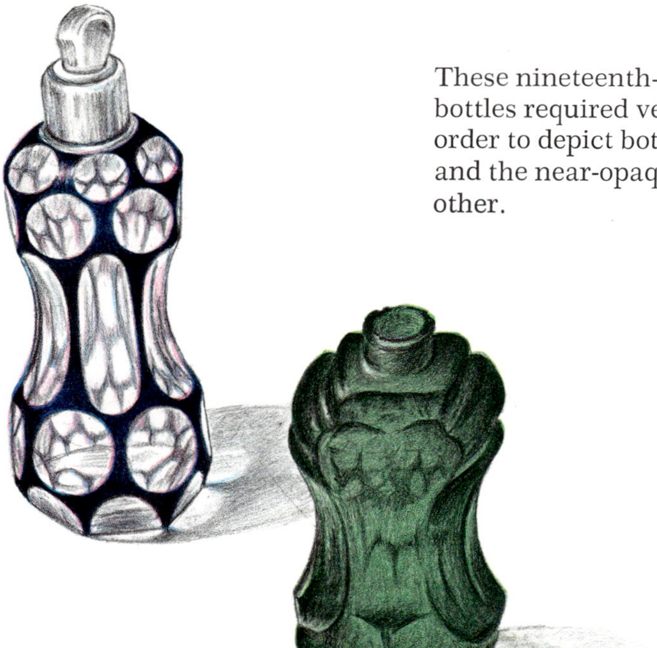

These nineteenth-century American perfume bottles required very careful rendition in order to depict both the transparency of one and the near-opaque "bottle green" of the other.

to illustrate a book on antiques that he was writing. At first I was somewhat doubtful because I had never tried preseparations for man-made objects and was not certain that my pencil technique would lend itself well to the rendition of materials such as metal, ceramics, glass, and wood. But a sample drawing reassured me; the

A combination of pewter and hand-painted ceramic in this old beer mug called for the use of all four printing colors in a wide range of tints. The warm gray of the pewter contains percentages of all colors.

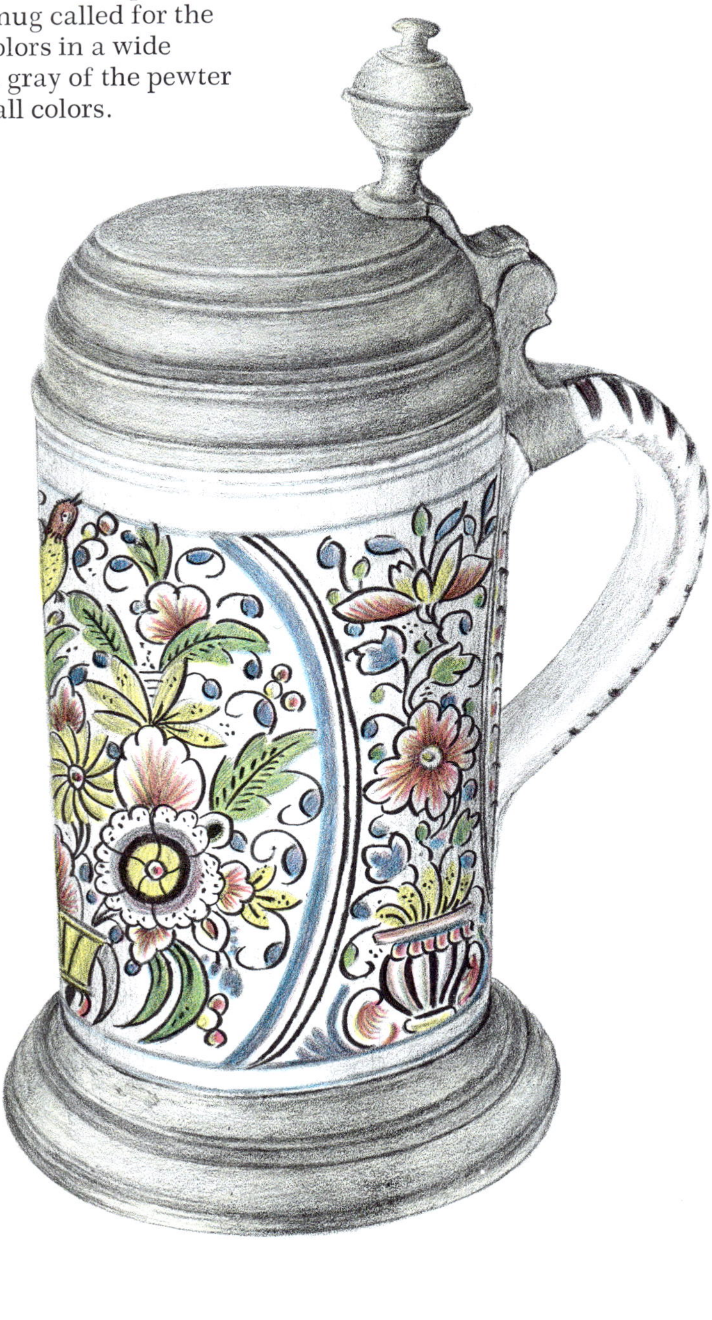

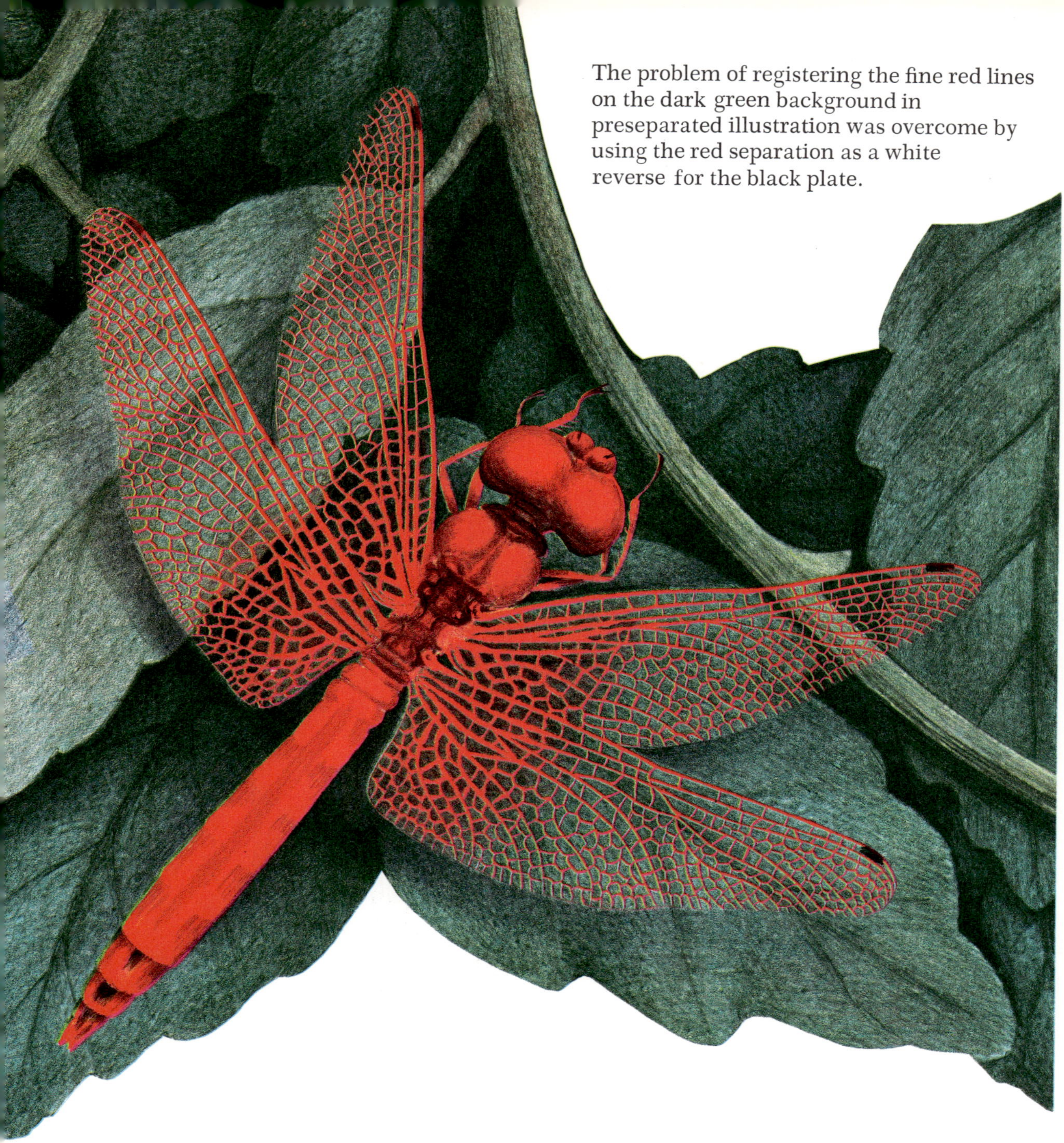

The problem of registering the fine red lines on the dark green background in preseparated illustration was overcome by using the red separation as a white reverse for the black plate.

pencil strokes could be adapted to reproducing each of these different textures with considerable fidelity, so I no longer hesitated to take on the job. The result was a handsome book, one on which the critics lavished praise for text and illustrations alike. One reviewer commented on the "incredibly three-dimensional" appearance of the objects depicted, adding that they all seemed to have a "special life and lustre." Another wrote that the illustrations seemed to make it possible for the reader to "feel the woodgrain or touch the glass." And so there was one more addition to my repertoire of preseparations.

Further experimentation resulted in some rather interesting techniques for achieving special color effects within the framework of the four separations. One was the occasional combination of a reverse and a positive image; another, the superimposition, in several colors, of the same image to produce specific hues otherwise not easily achieved, with perfect register at the same time. Such "tricks" can be very helpful in solving certain color problems and adding to the scope of a technique.

I may be conveying the impression that an extensive knowledge of color and its "anatomy" is crucial mainly for those engaged in preseparated artwork. Nothing could be further from the truth. Such knowledge is of great value to anyone involved with color reproduction in any capacity, regardless of the type of illustrations employed. Nor does it matter whether the separations are reproduced by photomechanical means or by an artist's eyes and hands. Appreciation of the basic rules of pigment mixtures and combinations permits a better understanding of both the possibilities and the limitations of color reproduction, and may on occasion even supply the means for a way out of some printing difficulty.

One such occasion arose during the printing of one of my early book jackets. It was a watercolor painting reproduced by the standard photomechanical process. As the first prints came off the press, it was obvious that the separation had not been successful. All the colors, from the greens and browns of foliage and branches to the shadows in the cluster of pink orchid blossoms and the insect pictured among them, had an unpleasant bluish cast that ruined the effect. Although clearly unacceptable as it stood, the reproduction would have required more time for the necessary color corrections than we had to spare. Seeking a way out of that dilemma, I studied the original and the separations side by side, and concluded that most of the excess blue could be neutralized by replacing the

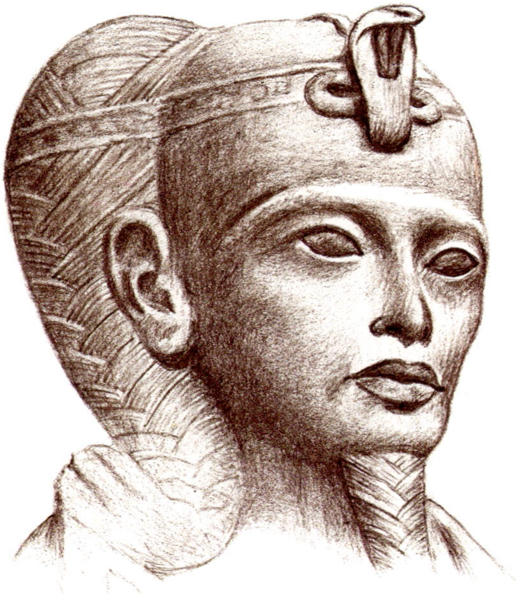

Black alone seemed suitable neither for this drawing of a sandstone sculpture of an Egyptian pharaoh, nor for the copy of a Dürer engraving on the right. Use of the same image for the yellow, red, and black plates produced soft brown.

magenta, or process red, with a red of a different shade. After calculating the exact percentage mixture needed to subdue the blue, I proffered my suggestion. At first everybody—and especially the printer—thought it an outlandish idea. Finally, however, they decided that the change was simple enough to give it a try. When it

By applying the same technique as above, the problem of registering fine lines on several overlays was resolved in this illustration of a feather.

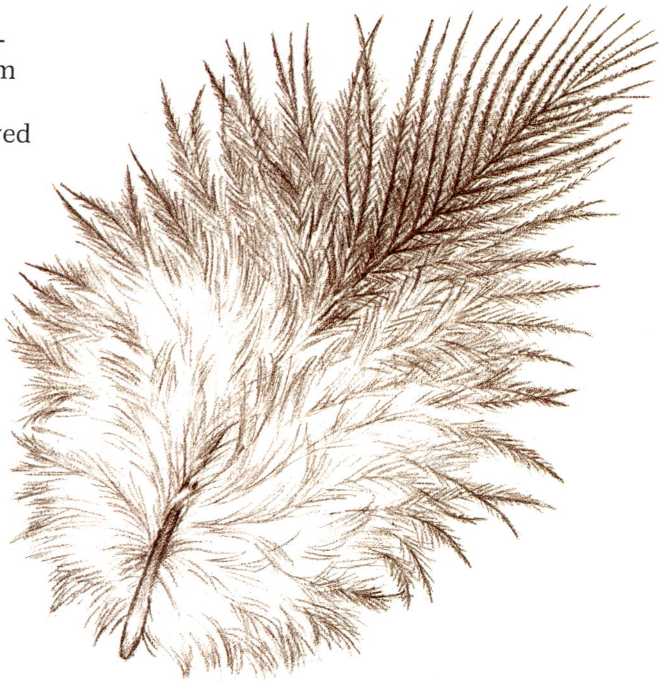

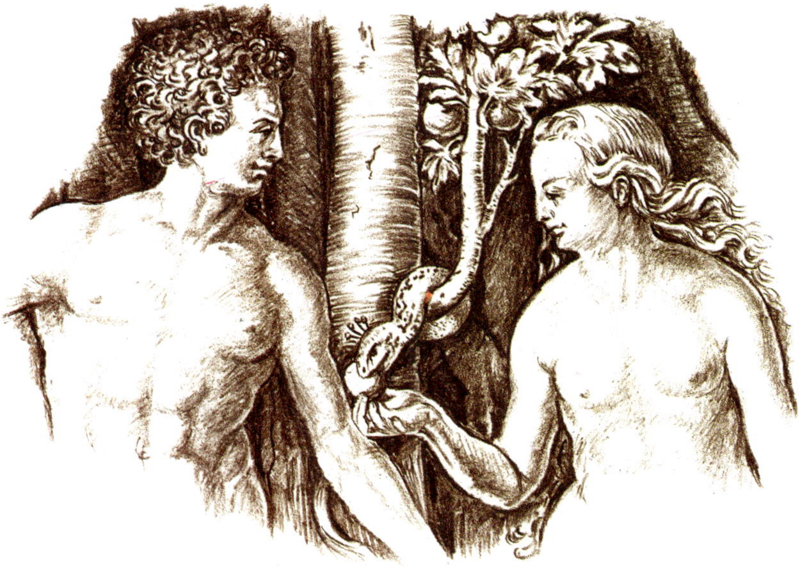

turned out very well indeed, I was the only one who was not surprised. Although the printed piece did not exactly reproduce the original's color values, it came close enough to be acceptable, and the printing was finished on schedule.

One of the greatest challenges in my experience with preseparated artwork was a book on iridescence, the "rainbow" hues that constantly shift and change because the slightest alteration in the angles of either light or observation causes different colors to be refracted. Iridescence is one of the most elusive and intriguing phenomena in the field of light and color; it was clear from the beginning that any attempt to illustrate such colors would be a somewhat frustrating experience. There is just no way for any image printed on a flat surface with pigment colors to yield a satisfactory reproduction of the nonpigmental iridescent hues. I did hope, however, that my illustrations might rival—and in some instances even surpass—the efforts of capturing iridescent colors by photomechanical methods, and I was not disappointed. Although not all the illustrations measured up to expectations, quite a few displayed a purity, brilliance, and translucence of color that exceeded those in most photomechanical reproductions. Another

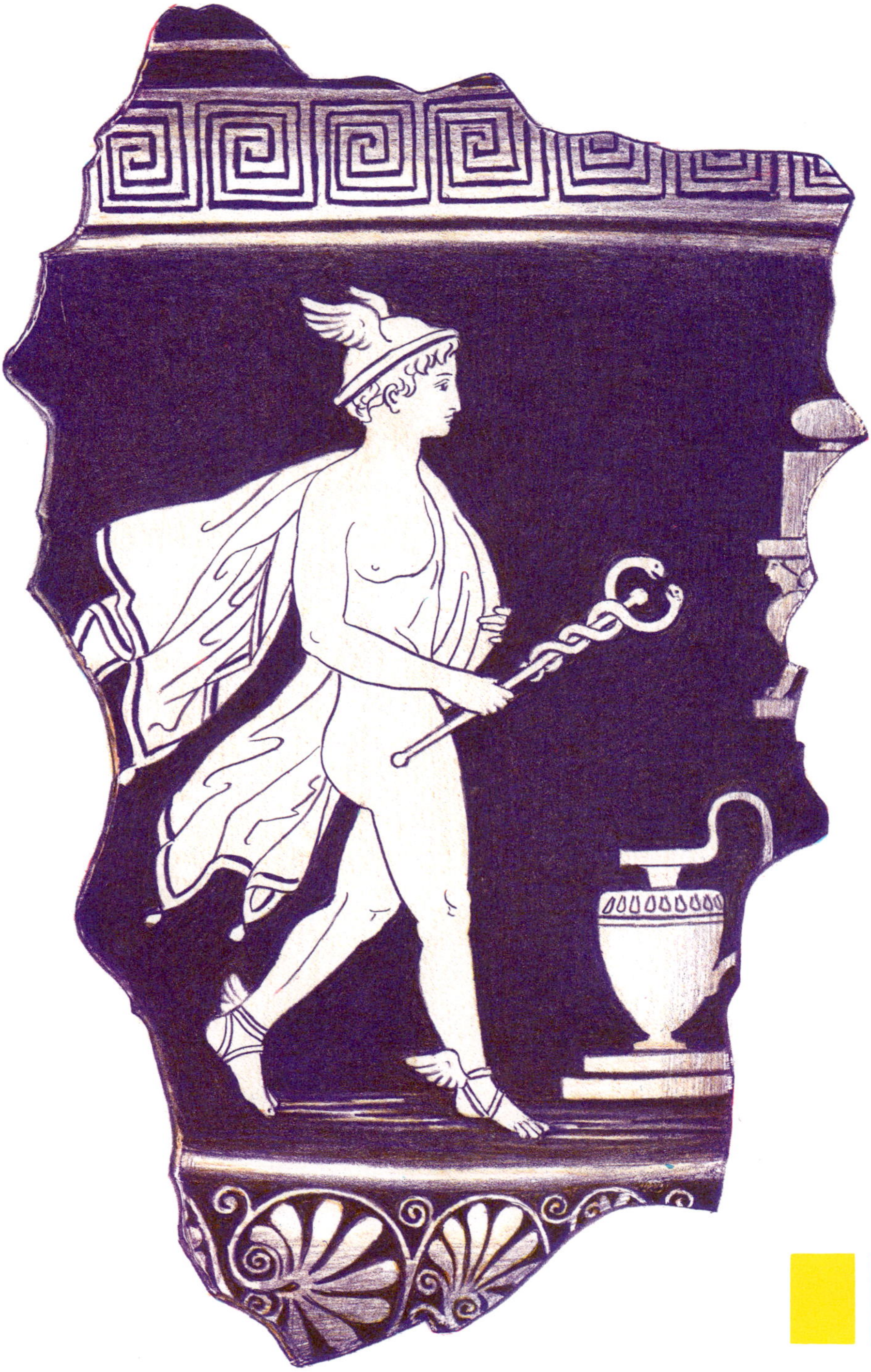

Using the same image for red and blue resulted in the handsome "royal purple" in this illustration of a Greek vase fragment. Black was not required at all.

addition to the spectrum of possibilities.

Part of the fascination of employing an unusual art medium lies in using it to create specific effects such as the softness of feathers, the luster of metallic colors, or the delicate texture of flower petals. A book on orchids with hues ranging from the palest to the most brilliant demonstrates the wide range of tints and shadings and the fineness of detail of which the halftone-simulating technique of preseparation is capable.

To me, one of its most satisfying features is that it reduces the mechanical part of the color reproduction process to a minimum. Contacted directly from the originals, the separations are not corrected or altered in any way by the printer either on the film or on the plate. The printed image is therefore as unadulterated as it possibly can be, and will look exactly as planned by the artist—provided, once again, that the printer is capable of doing his part of the job.

Although black alone would have been acceptable for this drawing of an archaeopteryx fossil, the blue-gray resulting from superimposing the same image in blue, red, and black yielded a more satisfactory slate color.

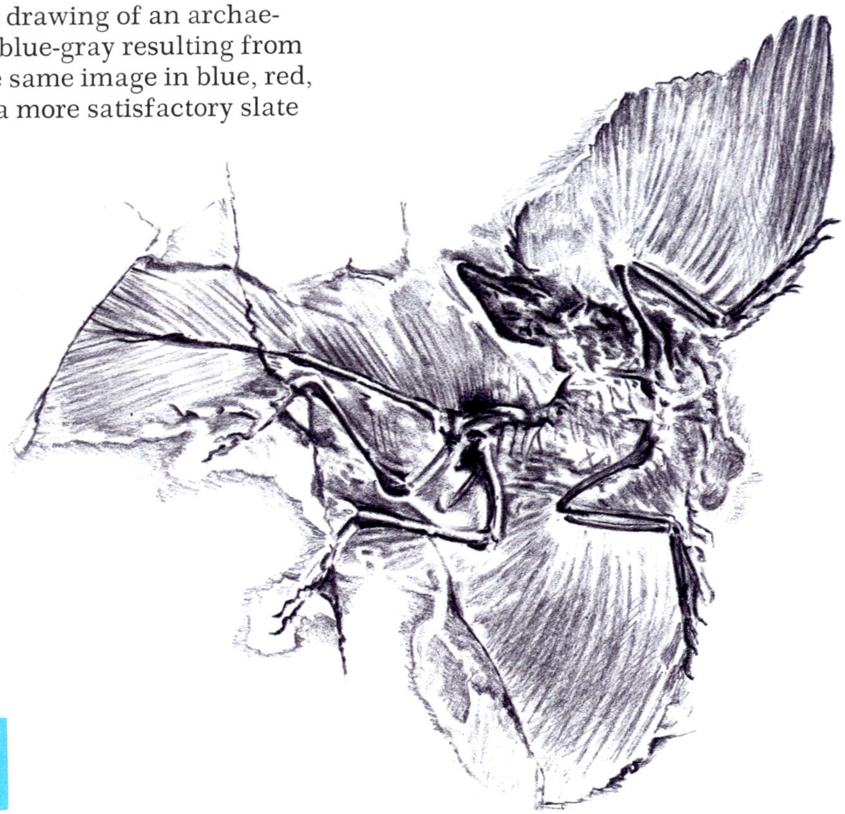

Compared to the photomechanical process, preseparated full-color art permits a great deal more flexibility in the choice of the four printing hues. For the reproduction of illustrations separated by the camera filter method, the three process colors magenta, yellow, and cyan are more or less mandatory. But since no filters are used for preseparated artwork, printing with the exact process hues is not a must—unless, of course, the full range of combinations that they offer is needed or desired. As an example, we will assume that violet-purple shades are not required for a specific subject, but that there is a need instead for a wide range of soft olive greens and browns, as well as for bright oranges and reds. Success in achieving delicate blendings of such hues would be greatly facilitated by replacing magenta with warm red, and cyan, or "process" blue,

Although iridescent effects can be only approximated by pigment colors, the butterfly shown here is proof that the ability to accentuate hues selectively in preseparations may yield a high degree of color purity.

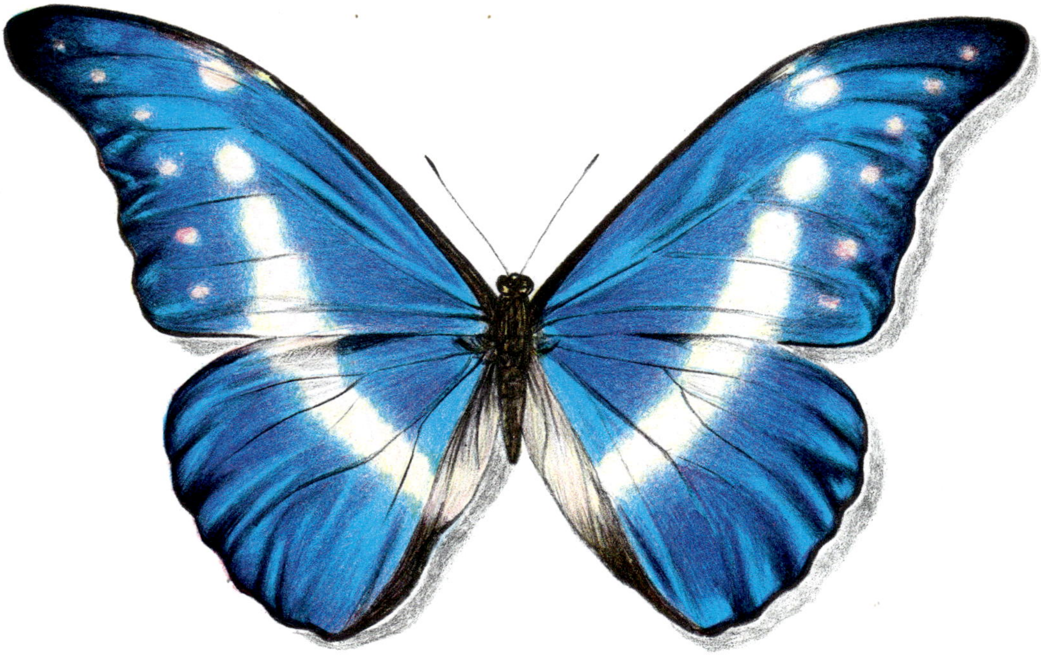

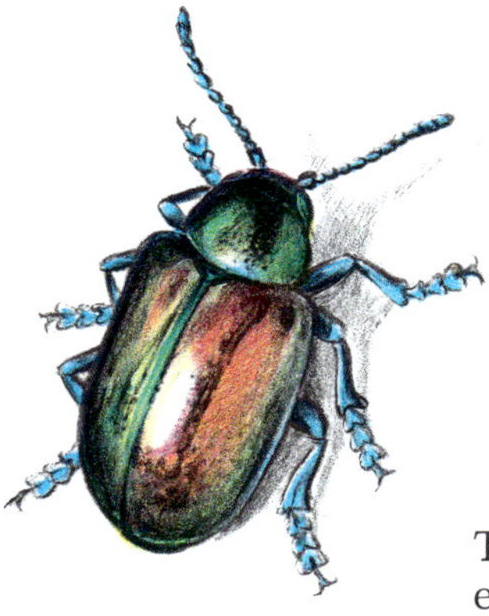

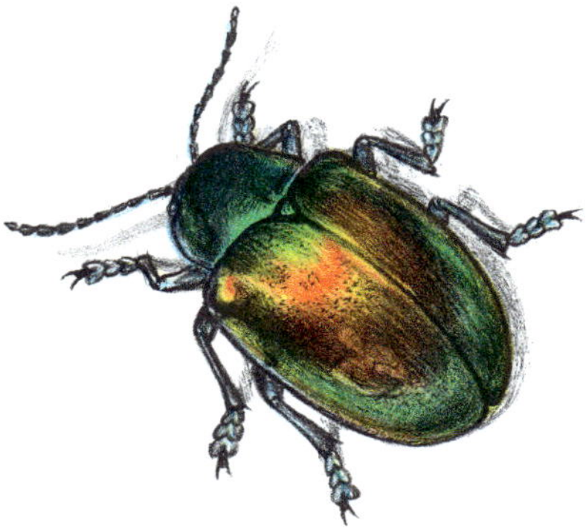

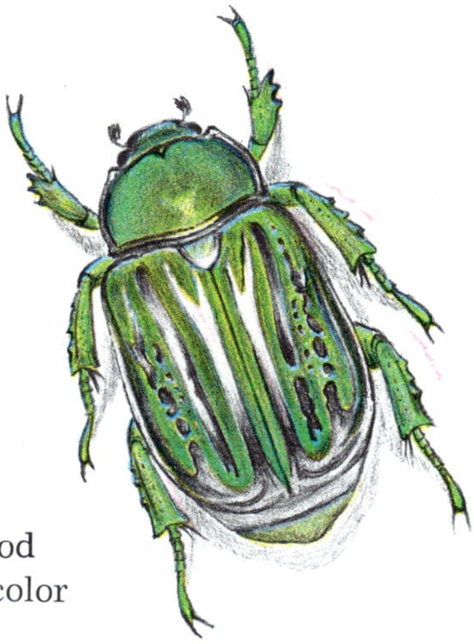

This reproduction of iridescence is a good example of the purity and brilliance of color that may be achieved by screenless pre-separation.

with a somewhat less greenish hue. Such color changes, of course, have to be reflected in all tints of the separations, and require a considerable knowledge of color. That, however, is true of the entire art of preseparation because it employs the mechanics of printing as an extension of a highly individual art form.

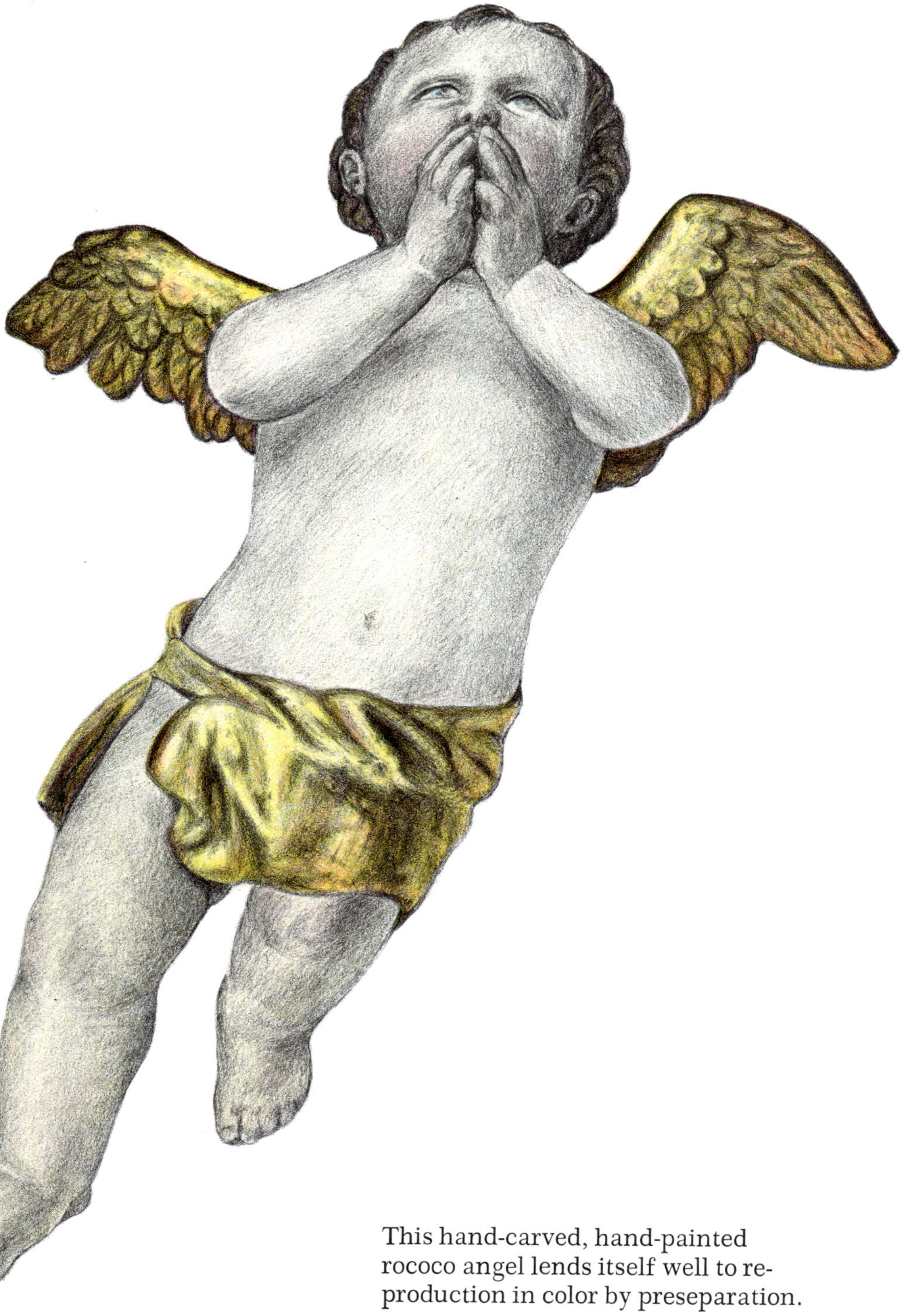

This hand-carved, hand-painted rococo angel lends itself well to reproduction in color by preseparation.

Epilogue

Living in a mechanized age, we have become used to machines and computers taking over an increasing portion of work previously done by human hands and minds. While mechanization has eliminated many boring and repetitious chores, it also holds a distinct danger, and one that is becoming steadily more evident to a growing number of people. For it has also led to dehumanization, to the turning of human beings into mere numbers on some punch card, and to the alienation and separation of people from the creative processes that once were inherent in many different types of work. This has resulted in the decline and virtual disappearance of the old-type craftsman, who with loving care studied and employed the materials and tools of his trade and took great personal pride in his work. The tremendous crowds at museum displays of ancient art and craftsmanship, be it in gold or other precious metals, in wood or pottery or weaving, clearly demonstrate the interest and longing with which today's users of mass-machine-produced articles look upon the creative, individualized products of the past. It is no coincidence that many of the old, half-forgotten "folk" arts and crafts from woodwork to needlepoint have experienced an unprecedented revival, finding new adherents who seek surcease from the relentless onslaught of mechanization in this creative work.

In illustrative reproduction the glut of slick mechanical pictures in books and magazines has led to satiation with this type of illustration. These often technically very good color photographs have become so commonplace that they are boring. They lack the individual appeal of a hand-rendered illustration that allows us to look at an object, a flower, or a bird not through the cold impersonal eye of the camera lens, but rather through the eye of the artist who beheld it and painted what he saw. There are, of course, many subjects for which the photograph with its objective, documentary fidelity is indispensable. But many times the photograph compares to the painted illustration the way the mass-produced piece of furniture compares to one lovingly and expertly hand-crafted by a master cabinetmaker.

For a while it seemed that hand-rendered illustrations would be restricted to juvenile publications. Recently, however, there has been a revival of interest in such artwork for certain adult publications. The Audubon Treasury *The Pleasure of Birds,* for example, features outstanding color photography plus two dozen handsome black-and-white illustrations, and several other recent books and articles featuring fantasy as well as nature also have been illustrated with watercolors and line drawings. This trend offers a chance to those artists willing to expend the necessary time, skill, and effort for the freedom to create quality illustrations in their own way, and thereby reintroduce individual and artistic dimensions to a field long monopolized by photomechanics. Comments by both readers and reviewers leave no doubt that many people respond to and appreciate creative illustration. Art techniques designed to yield good color reproductions therefore seem well worth the price in time and talent required to perfect and apply them.

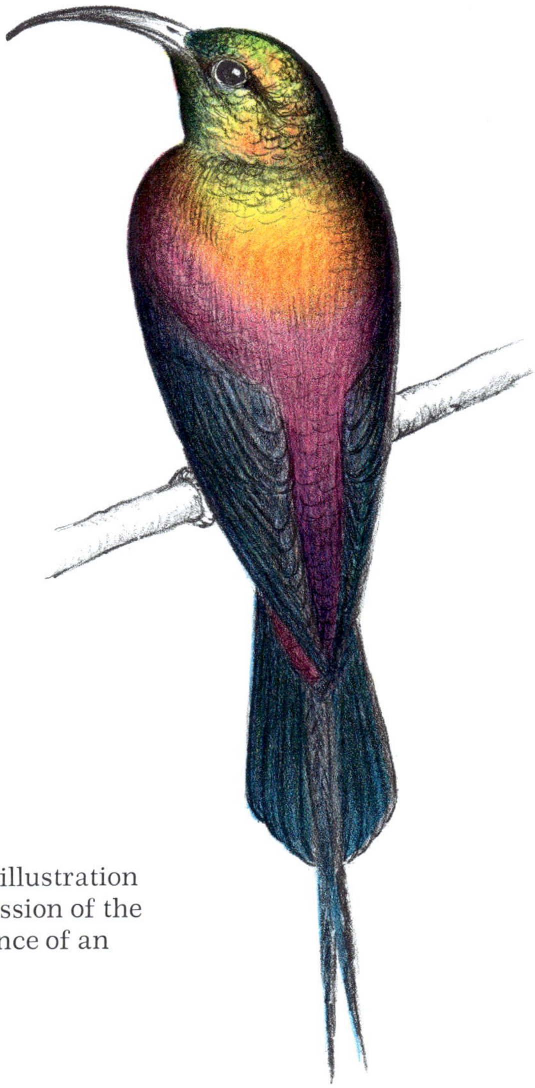

The preseparated colors of this illustration convey a reasonably true impression of the elusive "molten-metal" iridescence of an African sunbird.

Index of Names and Topics

Audubon, John James, 33

Bewick, Thomas, 32
black-and-white, 32, 35, 37, 39, 122
black,
  "dead," 37
  halftone, 36, 41–42
  ink, 27, 29
  key tints, 82
  lamp, 29
  photographic halftone, 49
  separation, 75, 93
  "skeleton," 98
  toned, 37
"bleeding," 66

"camera copy," 89
Chevreul, Michel, 76
"cloisonisme," 79
color,
  additive, 20
  charts, 80–87
  clashes, 78–79
  combinations, 72, 82–87
  complementary, 18–19, 20–21
  correction, 64
  filter, 25, 59–61, 63
  illusion, 74–77
  iridescent, 26, 115, 118, 123
  of light, 16–29
  of pigment, 16, 19, 20–25
  printing, 59–69
  process, 61
  secondary, 18–19
  sense, 13
  spectrum, 16–17
  stained-glass, 79
  structural, 26
  subtractive, 21
  transparency, 66
  vision, 13
  (*see also* illustration, separation)
compartmentalization, 80
complementaries, 18–19, 24, 61–62

Daguerre, Louis, 33
daguerreotype, 33
Da Vinci, Leonardo, 14
duotone, 41–42
Dürer, Albrecht, 14, 30–32, 73
  art techniques, 30–32
  copperplate engravings, 30
  woodcuts, 28

engraving, 30–34
  copperplate, 30–33
  metal, 33
  photomechanical, 33–34
  steel, 30, 32
  wood, 31, 32–34
etching, 32

filter, 25
  photographic, 59–61, 63, 69

Gobelins, 76
Gutenberg, Johannes, 30, 32

halftone,
  black, 33, 35, 36
  color, 56–66
  screens, 41, 45, 62, 75
Huygens, Christiaan, 15

illusions,
  color, 74–77
  optical, 74

illustration,
black-and-white, 34–38
full-color, 56–71
halftone, 33, 35–36
line, 34, 46
pen-and-ink, 34
pencil, 34–35
preseparated, 92–99
process, 56–68
techniques, 34–35, 43–44, 46
three-color, 48–55
two-color, 40–47
wash, 35, 47
watercolor, 35, 47

imposition, 66
"Impressionists," 76
ink,
black, 27, 29, 37–38
color, 36, 68
offset, 68
see also printers' ink
"intaglio," 30
iridescence, 26, 115

light,
corpuscular theory, 15
invisible, 16
long-wave, 16
primaries, 16–22
quantum theory, 15
reflection, 21–23, 25
secondaries, 18–19
short-wave, 16
spectrum, 16–18
transmission, 25
tri-chromatic theory, 17–19
wave theory, 15
lithography, 57, 59
offset, 59, 69
"luminism," 76

Michelangelo, 14
millimicron, 18
moiré pattern, 65

Newton, Sir Isaac, 15
Niepce, Joseph, 33

offset, 59, 69
optical illusion, 74

paper, printing, 68
photoengraving, 33–35
photographic halftones,
black, 33, 35, 36
color, 56–66
pigments, 14
Planck, Max, 15
"pointillists," 76
preseparation, 89–99
three-color, 48–55
two-color, 40–47
primaries,
fundamental, 16
light, 16–22
pigment, 16–25
psychological, 16
"printer," 61
printers' ink, 29
printing,
black-and-white, 29–35, 122
color, 36–37, 57, 59–66, 69
history, 29–36
process,
colors, 61
reproduction, 59 ff.
separation, 60, 65
progressives, 98

reflection, 21–23, 25
register, 65–66
Rembrandt, 30
Renaissance, 31

scanner, 65
screens, halftone, 41, 45, 62, 75
secondaries, 18, 19, 24
Senefelder, Aloys, 57

separation,
hand-rendered, 70–71, 92–97
photomechanical, 35–37, 60, 65
process, 60, 65
Seurat, Georges, 64
signature, 66
spectrum, 16–17

Talbot, William H. F., 33
Titian, 32

watercolor, 35, 47
wood engraving, 32, 34
woodcut, 28, 30, 32–33

Young-Helmholtz tri-chromatic theory of light, 17

Zeiss Optical Works, 60

Index of Illustrations

(Grouped by categories)

Art Techniques
copperplate engraving, 31
pen-and-ink, 34
pencil (halftone), 35
pencil (line) 34; *see also* Two-, Three-, and Full-color Illustrations
wash (watercolor), 35
wood engraving, 32, 33
woodcut, 28

Color Charts
two-color, 82–83, 86–87
three-color, 84–85, 86–87
four-color, 86–87

Diagrams

Light/Color Relationship
additive nature of light primaries, 20
color wheel, 24
fundamental color sensations, 12
light primaries, 18
pigment primaries, 19
reflection and transmission of colors, 25
selective reflection in colored light, 22–23
separation of white light, 17
subtractive nature of pigment primaries, 21
visible spectrum, 16

Color Combinations/Optical Illusions
clashing colors, 78
colors on light and dark backgrounds, 74
compartmentalization, 80
different but identical-looking hues, 77
four-color combination pattern, 72
identical but different-looking hues, 76

Color Reproduction
enlargement of halftone screen values, 75
photographic filters, 63
screen angles for process separations, 62

Full-color Illustrations

Photomechanical Process
artifacts, 56, 61
butterfly, 66
flowers, 60, 66
oil painting, 64

Screenless Preseparated
artifacts, 110, 111, 120
birds, 68–69, 102, 123
butterflies, 67, 88, 90, 118
flowers, 107, 108, 109
insects, 104, 112, 119
shells, 105
six-color preseparated, 58

Individual Separations
photographic process, 63
screenless line, 93, 95, 97, 99

Single-color (Black) Illustrations
copperplate engraving, 31
pen-and-ink, 34, 38
pencil, 34, 35, 38
photographic halftone, 33, 36
wash, 35
wood engraving, 32, 33
woodcut, 28

Three-color Illustrations
black drawing with flat screen colors, 52, 53
black halftone with hand-rendered colors, 50, 51
screenless drawing, 48, 54
three-color drawing printing as brown, 114, 115
three-color drawing printing as purple, 116
three-color drawing printing as slate blue, 117

Two-color Illustrations
black halftone with single-color screen tint, 43, 45
photomechanical duotone, 42
screenless drawing, 40, 46, 47